Quantum Energetics and Spirituality

Quantum Energetics and Spirituality

Aligning with Universal Consciousness

Volume 3

KENNETH SCHMITT

Copyright © 2021 by Kenneth Schmitt

All rights reserved. No part of this book may be reproduced or transmitted in any form or by any means, electronic or mechanical, including photocopying, recording, or by any information storage and retrieval system, without permission in writing from the author.

ISBN: 979-8-9851064-2-8

Kenneth Schmitt
Phone: +1-808-280-4041
Email: timeless1@twc.com
Website: https://www.ConsciousExpansion.org

Independently published by the author

Contents

Introduction   1

1. Significant Findings of the Quantum Sciences   3

2. Resolving our Limitations   43

3. Enhancing Spiritual Development   79

4. Aligning with Higher Consciousness   111

5. Realizing Our Personal Truth   149

6. Mastery of life   191

Introduction

This book is about using an understanding of quantum energetics to expand our conscious awareness into a higher dimension of resonant frequencies, taking our lives into a realm of love, joy, abundance and freedom. It begins with changing our emotional polarity from negative to positive, which may be a very challenging leap in consciousness.

I invite you to come with me on the inner journey to realization of our expanded Self. Examining life from a perspective of quantum energetics enables us to see through some of the distractions and deceptions that normally keep us enthralled. Once we become adept at feeling the quality of energy within and around us, we can choose wisely how to encounter it, and whether to align with it or transform our perception.

It is from within our own Being that we can know the truth about who we really are in our expanded Self and become the masters of our lives. The subject of this book is the quest to penetrate our limits (which are all self-imposed) and free ourselves to enjoy our unlimited awareness and creative abilities.

We can learn to be sensitive to our intuitive guidance, which is our connection with the source of our conscious life force within the universal consciousness that expresses itself throughout the quantum field and in every living being. This guidance has a qualitative energetic expression that is beyond the duality of our human experience.

We have lived within a limited compartment of consciousness as humans on this planet, but we are the only force that keeps us within these limits. As we learn to resolve our limiting beliefs and transcend our ego consciousness, we can align our thoughts and emotions with our intuition. This is the secret to living in a higher dimension of joy, love and sovereignty in the consciousness of the Creator.

Designed to be a guide for the inner journey of personal mastery and realization of the unconditional love inherent in our true Being, each essay presents a different aspect of understanding of our spiritual journey.

Kenneth Schmitt
April 4, 2022

1.

Significant Findings of the Quantum Sciences

Examining the Truth of our Being

If we really penetrate our inner knowing, we come to realize that our entire lives as human beings are illusory. For one thing, we believe that we live in a solid physical reality. For another, we believe that we are separate individual entities, apart from one another and from our Creator. A third illusion is that we are victims of circumstances, that everything that we experience happens to us from outside of our own consciousness, and that we affect our lives primarily by what we act upon and do.

From the experiments and conclusions of quantum physics, we know that the empirical world that we experience is a result of our conscious interactions with an energetic field of electromagnetic waves. There is no physical reality apart from our own recognition of the energetic patterns that we consciously imagine and observe. We live in a sea of energy that is an expression of consciousness. Our physical experience is a result of our conscious intent and interpretation of the energetic patterns that we attract by our own polarity and vibratory patterns, and that we recognize in the quantum field of all potentialities. There is nothing solid about our world. It is entirely composed of subatomic, spinning energy patterns that we perceive as material, and that stimulate our senses.

We cannot possibly exist as separate individuals, even though we believe that we are. This is a trick that we play on ourselves with our conscious compartmentalization. Quantum physics has shown that everything arises from consciousness, which is universal. There is only One consciousness, and it is the consciousness of the Being who creates everything. We arise from this consciousness and participate in it as completely as we allow ourselves to. From the reports of those who have died and returned to their bodies, as well as others who have expanded their awareness beyond time and space, we know that we are all of the same consciousness, which is eternal and pervades everyone and everything.

We are fractals of Creator consciousness, and in our true Self, we transcend time and space. We control our expressions as individuals in all of our interactions and experiences. We are completely free to create any experience that we desire with our conscious mental and emotional intent. Nothing happens to us that we have not created in quality of energy or opened ourselves to experience. It is the operations of our consciousness that create our life experiences. Once we realize that we have created our own illusory, limiting beliefs, we can learn to unlimit ourselves and expand into awareness of our true Being.

Understanding Our Virtual-Reality Empirical World

Since quantum physicists have shown that consciousness is the source of everything, cognitive scientists have begun to analyze our interface with the physical world and to understand how we perceive what we believe is real. Because our perceptions are subjective, scientists look to mathematics for understanding, because math is objective and indisputable. It is assisted by our understanding that everything is a fractal of its inherent design.

Nothing is solid. Everything that stimulates our senses is spinning patterns of electromagnetic energy waves held in form by Source consciousness, which we participate in, as fractals of Source consciousness. We have the ability to interface with the energetic patterns that we perceive as real. Our consciousness has no inherent limits. It is only our vibratory state of being that limits our awareness, and we have absolute control of this through our chosen focus of attention and the beliefs that we accept about ourselves.

Through the polarity and vibratory frequency of our state of being, we create the quality of our experiences. All of our experiences happen within our own consciousness. There is nothing outside of our own eternal Being. We envelope the entire cosmos of all electromagnetic energy in many different patterns and potentials. Each of us is a portion of Source consciousness, all being part of the same Being, and each having free will to determine our own quality of experience.

Through our focus on the thoughts and emotions that we entertain, we create the nature of our interface with the quantum field that we consciously inhabit. Thoughts and emotions are not part of our nature, but we become aware of them through our focus of attention and can experience them as our own. They exist in the plasma field of our environment. In every moment we have the choice of what we wish to be aware of, think about and feel.

Because we are part of the human community, we share our conscious interface with other humans, and we also have our own unique experiences. The quality of what we experience is not dependent upon others or our environment, only upon our own focus of attention. Circumstances arrange themselves according to our own perspective, because we are energy modulators. This is the essence of our creative Being.

Understanding the Basic Principles of Our Lives

From experiments in quantum physics, we know that there is universal consciousness that all sub-atomic wave/particles participate in. Photons, for example, can be in many places at the same time, and they know where they all are and what their status is, even if they're separated by light years apart. They are multi-dimensional and are not bound by space and time, but they can exist within space and time.

They are part of our physical and psychic structure, giving us their attributes of multi-dimensionality and participation in universal consciousness, able to live unbounded within space and time. Although we inhabit an empirical aspect of consciousness, we are not bound within it, even though, from within it, it appears as if we are. Since our bindings are all self-imposed in the form of limiting beliefs about ourselves, we have the ability to free ourselves in our awareness. Through our imagination and emotions, we can transcend our current circumstances and transform our lives into whatever we desire.

Universal consciousness is guided by an intelligence that we can call the Creator. Everything and everyone that exists is a fractalized expression of the Creator's consciousness. Since we have self-realization, we participate in creative ability by the operations of our psyche. Consciousness expresses itself in energetic patterns of electromagnetic waves. Our thoughts operate electrically, and our emotions operate magnetically. Both conjoin to form energetic patterns of polarity and frequency. Within the empirical spectrum, they manifest as experiences for us.

By our mental and emotional alignment with energetic pat-

Chapter 1. Significant Findings of the Quantum Sciences

terns, we create our life experiences. Our predominant thoughts and feelings radiate their energy into the quantum field of all potentialities and attract compatible energetic patterns that we recognize and bring into manifestation in our experience. It is our recognition that enables them to become real for us. We can control our experiences through the focus of our attention and calling forth our emotions.

As we learn how to choose and control our thoughts and emotions, we can control our state of being, resulting in the mastery of our life experiences. In this process we are guided in the most loving and joyous ways by our intuition, once we decide to become sensitive and aware of it. We can come into an understanding of how life works and how to master ourselves.

Enhancing and Enlivening Ourselves

We know from quantum physics experiments that universal consciousness is the source of all energy and material manifestation. The Being or Source of universal consciousness exists beyond the grasp of physics. All Beings who arise from universal consciousness also participate in the same consciousness. This is a logical truth. It is revealed in the awareness of masters who deeply understand consciousness. These are spiritual masters like Jesus, and some are on our planet now. There have always been some here to balance the energy of humanity. This has enhanced our flow of life force from the conscious plasma envelope of life surrounding everything.

In order to open ourselves to universal consciousness, we must resolve all of our self-imposed limitations. In our larger personal Being, we are unlimited in every way. In the traditional spectrum of humanity's vibrations, the human mind cannot comprehend this, but we can recognize that there's much more that we can experience. When we are open to it, our intuition brings realization to us through alignment with the energy of the heart of our Being. This is where we can experience the consciousness of the Creator Source of everything.

Our bodies have their own consciousness, which is guided by the vibrations of our ego consciousness. The ego operates outside the energy of the heart in the low-vibration realm of fear and mortality. We can withdraw our life force from the low-vibration spectrum of frequencies by paying attention to life-enhancing energy patterns. These are in the spectrum of love, compassion, forgiveness, gratitude, joy, personal sovereignty and abundance. We can connect intentionally with all of these vibrations, giving ourselves a perspective of compassionate wisdom. As we rise in our mental and emotional frequency spectrum, we expand our awareness and come to know the experience of greater love in our lives. Our bodies respond to these energies by regenerating and transforming into the manifestations of eternal enhanced awareness of beauty and health.

Self-Realization may come gradually or suddenly, depending upon our openness, conscious intent, and willingness to follow our intuition. It is in this direction that the flow of life-enhancing energy is growing brighter in alignment with the rising vibrations of the Earth into the spectrum of love and compassion.

Moving Through Levels of Consciousness

In the beginning of creation, we did not exist in our personal consciousness, but we were points of light in the universal consciousness of the Prime Creator. We were involved in the creative process beyond space and time. As our personal self-awareness came into existence, we could recognize ourselves as each an eternal presence of Being, always existing and always creating. We are pure awareness in the quantum field of unlimited energy forms and patterns. We are the modulators of energy, constantly creating by the vibrations of our thoughts and emotions. In our true Presence, we can express ourselves in any form and in any timeline and dimension that we choose to focus upon.

In order to have a real experience in the low vibratory realm of suffering, pain and fear, we agreed to have our awareness wiped clean upon incarnation on Earth. We have been unable to recognize our true Being and have experienced the full spectrum of fear. This is our final lifetime in a low-vibration earth human experience. We're here to clear our consciousness and guide humanity into a higher realm of open awareness of our participation in universal consciousness and Creator awareness.

As we come to elevate our awareness along with the rising resonant frequencies of our enveloping energetic flow, our personal relationships improve and become more heart-felt. Those who don't align with our energetic signature disappear from our experience. We are comfortable in higher-frequency living. Higher frequencies of thoughts and emotions bring less fear and more love into our lives.

To be truly clear in our consciousness, we must be able to be absolutely present conscious awareness all the time. This is what we are developing in our rising resonant energies. We are capable of awareness that may be infinite. There is so much available to us beyond the physical, and we have the opportunity to raise the vibrations of our physical bodies along with our consciousness. We are becoming aware of our ability to create experiences intentionally for ourselves, as they flow in harmony with unconditional love in universal consciousness. This is the electromagnetic direction that humanity is being drawn toward.

The Miracle of Compassion

Compassion is a high-frequency vibration. In our encounters with others, we always have a choice of how we want to feel and interact. Our best feelings are based in unconditional love, even when we are facing horrific evil. We cannot change the vibrations of others. We can only control our own vibrations, and we always have our own choice of how we want to feel. Our feelings control our vibrations and are broadcast into the quantum field enveloping us. They manifest our radiant aura, which is felt by everyone in our presence.

Compassion is different from sympathy, which is an alignment with the lower-frequency energy that we face, opening ourselves to feeding our life force to lower vibrations. Compassion arises from the love of our higher Being and is a transmission of high-frequency energy that lifts the vibrations that we encounter into alignment with the attractive energy of love. We can know compassion by our alignment with the energy of the heart of our Being. By feeling and knowing intuitively in every moment the promptings from our innermost Being, we can be true and expansive in every encounter.

Our innocence is our protection from low-frequency attack. By living in compassion and love, we are in a higher dimension than low-frequency energy attacks and are invisible to them, unless we choose to encounter them by lowering our vibrations in sympathy and fear.

In order to have true compassion and love, we must resolve all of our limiting beliefs about ourselves and know that we are eternally conscious Beings of unlimited creative power. This is who we are beyond our limiting human experience, and we can come to know our true identity while embodied on this planet. We are of the same essence as the universally conscious Self-Realized Creator of all, whose consciousness we arise out of and participate in. We are fractals of the One. Based on what we know from quantum physics, this is the only logical reality, and it is the experience of all spiritual masters. This high-frequency level of experience is available to us by our choice. We only need to develop our inner sensitivity and awareness of our intuitive knowing, which has been dormant for eons of dense material living as humans. We can call it forth in our consciousness and open ourselves to miraculous lives.

Chapter 1. Significant Findings of the Quantum Sciences

Knowing What's True

For eons we have regarded our mental processes and our will power as the source of our guidance and success in life. We have imagined that this is our inner power and the closest we can come to knowing the truth about life. We have limited ourselves to empirical experience and logic. Higher math comes closest to true knowing in the unenlightened mind, because it depends upon intuition. Only in the discoveries of quantum physics have we learned that there is much more to understand.

We have learned that in our essence, we are our conscious awareness, which is unlimited beyond time and space. In our personal awareness, we are eternal, just as universal consciousness, which we participate in, is the eternal awareness of the Creator of all that exists. We are all fractals of the One Universal Being. There is only one consciousness, and we are designed to be able to have access to all of it. Our consciousness envelopes the entire cosmos and all that is.

We can find ourselves at the beginning of our own awareness. This is the state of being where we just are present awareness. This awareness is our conscious life force flowing into us as unconditional love, connected in Being through universal consciousness with all conscious beings, which is everyone and everything. Everything has its own level of consciousness constantly flowing into its form and presence, with its own energy signature. Life flows through everything, expressing itself as energetic patterns of electromagnetic waves, some of which we can perceive as empirical experiences.

We exist in many dimensions simultaneously. Everything that exists is present to our awareness here and now. We are aware of as much as our personally-created limitations will allow. We have had many beliefs about ourselves that have kept us compartmentalized in our awareness. We don't need any of them, but they have been with us nearly our entire lives. They are all based on the fears and needs of our ego consciousness, which knows nothing beyond empirical experience and mental concepts, although we do have our imagination, which can be a link to higher awareness.

The ego, however, cannot bridge the gap beyond mortality in time and space. We can thank our ego consciousness for giving us our human experiences in the empirical dimension, while we open our awareness to our etheric presence in a dimension of higher vibrations. In our etheric presence we can recognize our eternal Self.

Awareness beyond Belief

Quantum physics has proved that there is a universal consciousness that is the source of everything. Anyone can conduct these experiments and get the same result. This makes universal consciousness a scientific law. Everything has a consciousness that enables it to exist as its unique expression of being. What is our unique expression of being? How much conscious awareness are we capable of?

Let's examine our limitations and determine how to expand beyond them. When we incarnated, we lost our sense of Selfhood. We became dependent upon others for our physical nourishment and well-being, as well as learning the ways of human interactions. We developed ego consciousness and adopted an awareness of separation from the source of our Being. We learned to live in fear of termination. We have adopted a long list of beliefs that keep our awareness compartmentalized into a spectrum of vibrations that we all share. We believe in a physical world of reality outside of our own selves.

We know that there are mysteries that don't fit our version of reality. People have died and come back with amazing accounts of greatly expanded awareness. Some people are able to levitate, move objects with thought, know the thoughts and feelings of others and engage in conscious travel beyond the body. Some of us with disciplined minds and emotions can create things and change our circumstances at will.

When we learn to meditate deeply and penetrate our consciousness beyond our beliefs, we are able to engage with the patterns of unseen energetic vibrations that envelop us in the quantum field. We can enter dimensions of frequencies beyond our human experience. The higher we go, the more beautiful and loving we feel. We can feel into an ecstatic realm in the vibrations that come through the heart of our Being. This is where we can know everything without thinking. We become pure personal awareness, unlimited in every way.

With practice we can maintain awareness of conscious unconditional love flowing through us and radiating our inner light of photons into the quantum field that our awareness envelops, attracting all energies and beings around us into alignment with our energy signature. This is how miracles are created.

Choosing our Personal Circumstances

We are the creators of our personal reality, and we also influence everything beyond our own being, because we manifest ourselves as energetic presence. The most significant aspect of our being is our energetic vibration and polarity. We manifest ourselves as either negative, fear-based persons or positive, love-based persons. We can also be neutral observers. It is the neutral place where we can best detect the quality of energy in our presence and choose to align with it, just observe it or transform it.

We become aware of the quality of the energies in our own being and in our encounters through our emotions. When we feel unhappy and unfulfilled, we are in low-frequency, negative and life-draining energy patterns. When we feel happy and joyful, we are in high-frequency, positive, light-filled and life-enhancing energy patterns.

We can feel that we have no choice about our experiences, especially when our circumstances are deeply negative, but this is not true. Through our attention, consciously and subconsciously, we have created the quality of our circumstances. Through the vibrations that we choose to hold in our thoughts and feelings, we electromagnetically attract the same quality of vibratory patterns into our experience. It is the intersection of our predominant thoughts and feelings that creates our personal energy signature. This is the frequency that we radiate into the quantum field for manifestation into our experiences.

We are designed to align with the positive, unconditional love of the universal consciousness of our Creator essence. This is the direction of the flow of our conscious life force and is our natural state of being. This is the reason we feel good when we're in the love vibration, and we feel bad when we're in the fear vibration. We're capable of many complex thoughts and feelings and personal dramas, but if we can understand the simple structure of our essence, we can choose to direct our lives in the best ways we desire.

Realizing Our Innate Abilities

We have lived in the low vibrations of victimhood, suffering and fear for eons. We have always wanted a better life of freedom, love, peace, beauty and abundance. Let's examine the energy that has held us captive in experiences that we do not enjoy. We have blamed our oppressors for our suffering, but are they really the cause of how their actions affect us?

To determine the truth of our experiences, we can examine the electromagnetic energy patterns involved in our lives and how we interact with them. Quantum physics has taught us that everything we experience is energy. By recognizing patterns of polarity and frequency of electromagnetic waves in the quantum field, we cause them to materialize in our experience. They

Chapter 1. Significant Findings of the Quantum Sciences

appear as we imagine them to be within our belief system. It is our recognition and awareness that creates our experiences.

What happens when we face threatening situations? Every scenario in our lives is a complex combination of energetic patterns that manifest for us, because we recognize them and believe they are real. Does that mean that if we were arrested unjustly and imprisoned, that we could adjust our focus so completely that we would recognize the quality of the situation, give it no life force from our attention, and just disappear from the scene into a new situation that we have just envisioned? Yes, if we can be so powerfully focused in our imagination and emotions that we have absolute confidence about what we're doing, and we are totally, and in full confidence, involved mentally and emotionally in the new scenario, without any emotional alignment with the current physical situation.

Quantum physicists have determined that a universal consciousness is the source of all of the energy in the unified quantum field, which provides the essence of everything that exists in all dimensions. There is only One consciousness and One conscious Being. Each of us is this Being in the sense that we are fractals of universal consciousness. Our living essence of Being is creative in every way. We can modulate the energetic patterns in our awareness according to how we feel and what goes through our mind.

All energetic patterns are attracted and repelled by the polarity and frequency of our perspective on life. This is the vibratory level that we are predominantly aware of. When we are aware of the subtleties that we are capable of emotionally, we can let our feelings tell us everything we want to know about the energy in our environment. In any situation we can summon our emotions to align with any vibration that we choose. Our emotions then bring our thoughts into vibratory alignment as much as we permit. We are the ultimate designers and producers of our lives.

Penetrating the Heart of Our Being

Our physical heart has its own neurological system and is the most powerful organ in our body. It has 60 times the electrical power of our brain and 5,000 times more electromagnetic influence than our brain's will power. Its magnetic field has a toroidal shape that extends out several feet from our physical body. It can operate in coherence with our brain and is the conduit of our intuition and creativity. Ethericaly our heart is the central source of conscious life force that flows from the universal consciousness of the Creator.

By focusing on our breathing, taking deep, rhythmic inhalations and exhalations, and being attentive to the rhythm of our heart, we can calm our mind and emotions and attune to our intuitive knowing. We can learn to clear ourselves of thoughts and matters of ego consciousness and just be present in awareness of the energy of our heart. It's helpful to be in a peaceful area of nature without human influence or in a quiet room, perhaps with soothing music.

We can focus on a point of light within the consciousness of our heart. This is the manifestation of our conscious life stream and the unconditional love of our Creator consciousness. It emits photons, tiny quanta of light, that create our radiant aura. It is our connection to the unified quantum field of all potentialities. It is our access to all the energy patterns that we can encounter, and it is where we modulate the patterns of electromagnetic waves that we choose to recognize and feel in the quantum field. This is our point of magnetic attraction to all of the energy that resonates in alignment with our personal energetic signature. It is how we create the quality of our experiences.

By choosing to have a positive polarity with high-vibrational thoughts and feelings, we create experiences that best align with the energy of our heart. It is the energy of great vitality, joy and the feeling of connection with the essence of all conscious beings. It is how we can become en-light-ened in realization of our true Being and is the source of our power to change the world.

Creating More Meaningful Lives

In the observations of quantum physicists, they have determined that there is only one universal consciousness that everything participates in, from the smallest sub-atomic particles in the structure of our world. This is the consciousness that we participate in, but we can access only as much of it as our vibratory level allows. Historically humanity has also been constrained by the energy grid imposed by negatively-polarized beings who controlled us. This grid has now been disabled, and the beings are no longer present. We are free to evolve without constraints, and the rising resonant frequencies of Gaia and our enveloping quantum field are prompting us to expand our conscious awareness.

Conscious expansion occurs by opening our awareness to our inner knowing, our intuitive guidance. Most of humanity has been desensitized to our intuition and higher guidance and must be awakened to our potential. Because all of life operates energetically, we must learn how to interact creatively with the electromagnetic wave patterns that envelope and interpenetrate us. These occur in octaves of resonance, some of which we can perceive as humans. The empirical world has higher octaves of vibratory patterns that we are not presently aware of, but they are present within and around us.

As we open ourselves to higher frequencies of positive polarity, we can feel love and joy arise from the heart of our Being. When this happens, we can know that we are aligning ourselves with the unconditional love of the Creator consciousness. We constantly receive intuitive promptings offered by our higher Self, our soul consciousness. We can train ourselves to be aware of these promptings, which can be of many different kinds of sights, feelings, synergies, words and more. All of nature communicates within the spectrum of life constantly, and we can align with this energy. If we go deeply within, we can intuitively know the essence of every being and circumstance.

We can be in gratitude in every moment, filled with awareness of the infinite beauty and majesty of our true Selves. We can be completely Self-motivated, as we align ourselves with our intuitive knowing in positive, high-vibration thoughts and feelings, able to create whatever we desire for the highest good of all concerned. We are not mere humans. We are extensions of the consciousness of the Creator, fractals of the divine with unlimited abilities. We only need to open and free our awareness and intend to feel and imagine the most wonderful energy patterns that can manifest out of the quantum field as our experiences.

The Importance of Joy

When we are joyful, we elevate ourselves beyond the reach of negative energy. Life becomes a game in which we recognize low-vibration persons and situations from a perspective of love and compassion, transforming our encounters into alignment with us. If they can't align with us, we can just change our focus of attention to more fulfilling situations and withdraw our life force from the fear-tinged vibrations. Low-frequency people and scenarios cannot invade our presence, unless we allow it consciously or subconsciously. We can choose mindfulness and objectivity and protection in the higher vibrations of light and love.

Because the cosmos consists entirely of patterns of electromagnetic waves in the unified quantum field, our experiences depend entirely on our vibratory level. This understanding of the nature of our reality was recognized by quantum physicists and geniuses like Nicola Tesla and Einstein. Einstein didn't understand it to the depth that Tesla did. To Einstein the quantum field beyond time and space, in which the energetic expression of particles can be in more than one place concurrently, was spooky. Tesla understood the quality of energetic frequencies and helped to bring science into the realm of esoteric spirituality.

As participants in the energetic spectrum of humanity, we have developed numerous beliefs that have bound us to a world of low vibrations, living in constant expectation and fear of diminishing life. We are not required to stay in this perspective. We have free will to vibrate at any level we choose, regardless of what we recognize outside of our body. In fact, what we recognize as real outside of our body is actually a reflection of the energies within our own consciousness. Among us, the adventurers in conscious exploration have discovered some of these things apart from knowing quantum physics and spirituality.

We are real Beings beyond anything that we can imagine as humans, because we created our ego consciousness to keep us from knowing our true essential Being; otherwise, we couldn't be fully human in this spectrum of energy. We are, however, transitioning into a higher dimension of frequencies, as we can recognize from the rising resonance of the Earth and our surrounding galactic environment. To align with these vibrations, we are being prompted to be in joy and compassion. As we withdraw our recognition of their reality, the lower-frequency, negative vibrations fade out of our experience, and we no longer align with their energetic level. We are here now to awaken to a new reality and manifest it in our lives through joy.

The Essence of Our Being

Quantum physics has proved that all parts of our bodies participate in the universal consciousness that expresses itself as the unified quantum field of all potentialities. Within this field, everything exists, down to sub-Planck wave patterns and particles and beyond our current comprehension. Everything is energy, and all energies are expressions of universal consciousness. Deeper than this, science has not been able to go yet. Physics has already gone into quantum realms beyond the empirical, indicating that our current awareness can expand into other dimensions. Without emotional involvement in scientific observation and mathematical measurement, scientists cannot know the polarity and quality of patterns of energy. Yet, our intuition can go far beyond scientific knowledge. It is the source of genius for all who can align with it.

All of us can know the feeling of negative or positive polarity instantly when we encounter it. In their essence these energies feel either life-diminishing based in fear, or life-enhancing and empowering in love. In every moment we make the choice of which of these polarities we want to align with. They are mutually repellent of each other. If we try to accommodate both, we become disoriented in our judgment. The wave patterns of our focus of attention are not in resonance with each other. They can never be in resonance in our presence, unless we give them our life force through our attention and energetic engagement; otherwise, they disappear from each other's presence.

Our emotions express the quality of any energy pattern that we imagine or encounter. We know when we're in the presence of the vibratory pattern of love, compassion, gratitude and joy. This is the positive polarity of life. The opposite happens when we align with negative polarity. It is the realm of everything that diminishes life everywhere in our awareness. It can manifest in our world only from our life force through our recognition and engagement with it, because it receives none from universal consciousness, since its essence is of negative polarity.

In every moment we choose our polarity and our vibratory frequency. This choice determines the quality of our experiences. Everything that the quantum field manifests for us moment-to-moment is a result of our mental design and emotional state. These are our energetic modulators and give us our potentially unlimited creative abilities.

Creator Consciousness

The nature of the prime Creator, God, is also our nature. We know from quantum physics experiments and extrapolations that there is a universal consciousness that contains within itself everything that exists. Everything is a fractal of the Creator, and as the creator changes, all the fractals change exactly the same way in resonant energetic patterns, just like in an Escher painting, where every image is a fractal of the whole, and they all change mathematically in perfect interlocking patterns. Fractals aren't just mental, however, they are also emotional and intuitive. We are all designed to move innately with the energetic expressions of the Creator and be guided by our intuition, which is our connection with Creator consciousness.

Universal consciousness is pure awareness and knowing. It expresses itself by multiplying itself in order to have experiences of encounters. Each of us is the Creator, able to be aware of the higher truth in every situation. We create the qualities of our experiences just with our state of being—our thoughts and feelings. We can change the level of vibrations in our encounters by how we react to them. We can maintain a positive polarity with gratitude and joy. This is a life-enhancing state of being. As we move further along the path toward expanded consciousness and life enhancement, we more closely align with the natural energy of every aspect of life. This is Creator consciousness in expression.

We are the Creator's expression of personhood. Universal consciousness is not personal, it is universal. Through us, in our current state of being, the Creator can experience our personal identity and ego consciousness. We are supported in every way that we choose to go. Universal consciousness is not judgmental. All intentions are accepted, and guidance is always offered. If we choose to live in a positive polarity, high-vibrational state of being, we are supported in every way, because this is the intention of creative energy.

In every encounter we have an immediate emotional message of vibrational level and polarity. We can enhance and change the polarity and vibratory patterns we encounter with our state of being, our thoughts and feelings. By imagining and feeling the higher octave and positive polarity of every negative situation in this world, we can elevate the state of humanity, as well as ourselves. This is the most powerful way to change things.

How Far Are We Willing to Go?

Everything that we know about, in its essence, is an electromagnetic wave pattern. We are enveloped by infinite numbers of them comprising a quantum field of all potentialities. As conscious Beings, we express ourselves electromagnetically and are able to interact with any of these wave patterns. We have our conscious magnetic and electrical polarities that attract wave patterns that are in alignment and repel those that are dissonant. Our polarity is either positive, life-enhancing or negative life-diminishing, in alignment with our dominant mental and emotional states.

The farther we go in maintaining a positive, life-enhancing perspective, the greater our awareness becomes. We can transcend our limiting beliefs, which we can recognize as unnatural, fear-based and of negative polarity. We needed them for our human experience, but now they are irrelevant. They don't feel comfortable, because they don't align with our natural state of Being, which is sovereign and free. All of our limitations are self-imposed, and we can unlimit ourselves by resolving every limitation that we become aware of, or by transcendence in consciousness. This requires focusing only on vibratory patterns that are positive and of high vibrations. It is the realm of compassion and unlimited, greatest love.

It is a leap in consciousness for us to become positively polarized. It is the move from a perspective based in fear to one based in joy. We decide in each moment which polarity we align with. This is usually a subconscious operation, and requires training to change it. We can begin the process by aligning with positive, high-vibration feeling and envisioning. We can imagine scenarios with joyous, loving people having wonderful and exciting experiences. This elevates our energy signature toward those vibrations and attracts compatible ones. It also introduces our subconscious to those vibrations. With practice the old limitations disappear.

With intentional focus on positive, good-feeling thoughts of gratitude, kindness and feeling connected in our essential Being, we can elevate our energetic signature, making it possible for us to recognize our eternal present awareness beyond time and space.

Quantum Mechanics and Spirituality

Our minds operate with electrical charges, frequencies, amplitudes and polarity. Our emotions operate with magnetic polarity and wave patterns as well. They both intersect at 90-degree angles of each other, and they flow together always, forming a single wave pattern. We have the ability to recognize electrical wave patterns as thoughts and to feel magnetic polarity and vibration. We can also change the wave patterns by refocusing our attention. We can vibrate at a different level than the energy patterns around and in front of us. Through our attention, we have the ability to modulate the wave patterns that we become aware of with our thoughts and feelings.

By being compassionate and wise, we can encounter any situation from a positive, high-vibratory perspective of kindness, compassion and love. We can relate to the light in everyone and everything. In the energy of our heart, we all share in the consciousness of the Creator. We play different roles, which we have chosen for ourselves in order to have certain kinds of experiences. When we encounter one another, we always have the choice of a perspective with a polarity and frequency.

Because of the polarity of our thoughts and feelings, we attract compatible energy patterns and repel energy patterns that are of opposite polarity. Positive polarities are in alignment with each other, as are negative polarities with each other. We attract energy patterns and experiences that resonate with our own energy signatures and states of being. We attract experiences that resonate with us.

On the path to inner knowing, we can resolve all of our personal drama, as we proceed in compassionate wisdom. We can arrive at a place of present awareness, without thoughts or feelings, just being and knowing. We can train our attention to be present and aware. From this state, we can create the quality of life that we truly desire. We can feel and know the polarity and frequency of the wave patterns that we create.

As we develop sensitivity to our intuitive knowing, we gain awareness of higher guidance in every aspect of life, while having free will in every moment. This moves us into expanded awareness and more fulfilling thoughts and emotions. As we increase our sensitivity to subtle vibrations, we begin to enter the consciousness of all beings and everything. It is the consciousness of the Creator, imbued with unconditional love and brightness.

Directing Ourselves to Our Greatest Desire

We are all influenced by the auras of the energies around us, especially the resonant frequencies of the Earth and of the beings around us. Yet, we have free will to choose the quality of vibrations that we focus upon in each moment. With a strong intention, we can transcend all the distractions around us, many of which are designed to take our attention away from our true Being. We can know the emotions of our greater Self, all the feelings imbued with goodness and joy, compassion and gratitude.

Chapter 1. Significant Findings of the Quantum Sciences

We can know our innate wisdom, resulting from the experiences of many lifetimes. All of this is beyond our ego-self, but we can know these energetics through our intuition, both emotionally and imaginatively.

It may help our understanding by realizing the nature of our lives in the traditional human dimension of energetics. We have lived within the confines of vibratory frequencies of negative polarity, all of them based in fear of suffering and termination of ourselves. This is the realm of ego-consciousness, enclosed within the spectrum of fear. To change this, we can change our polarity to positive by intentionally reversing the polarity of every situation we are in, as well as our own perspective, to a vibratory level based in life-enhancing feelings and thoughts for ourselves and all other beings.

When we are consistently living with a perspective of life-enhancing energies, we attract resonant frequencies and patterns, helping us to sustain our perspective and empowering us. We can align with rising vibratory patterns. By living this way, we open ourselves to a greater flow of conscious life force, expanding our awareness within a greater consciousness.

Everything that exists is a fractal within the consciousness of the Creator. With mathematical precision, it all changes for us according to the constantly-changing vibratory resonance that we hold, as our fractal of Self-identity. We are the change-makers of all the energetic patterns around us. Every situation that we face is a reflection of the quality of energies in our own awareness, and we have the freedom of imagination and emotion to shape them as we desire with our conscious life force.

Through our experience here, we have developed many dysfunctional emotional patterns and thoughts. As we are able to resolve the causes of these, we can align ourselves more clearly with the polarity and resonant frequency of the Earth. This is why mystics have often chosen to live in wilderness areas. For us, though, the challenge is to maintain a natural perspective while living in society.

Just like in MC Escher's graphic designs, changes happen in increments, and we adjust to them. At times, however, there is a complete pattern change in order to maintain the integrity of the whole. An instance like this would be a change in our personal polarity, when we become naturally good people, with intentional energetics that reach deeply into our consciousness, and apparent miracles can occur in order to keep us in alignment with our chosen energetics.

Creating Our Energetic Presence

Each of us has become an energetic presence of our belief structures. If we change our beliefs or just resolve them, we become an energetic presence of our new state of being, and our lives change according to our vibratory level. Energetic patterns manifest as we recognize them and feel them to be, either in our imagination or in our physical experience. What manifests is the vibratory quality of our state of being in every moment.

With determination, each of us can have a positive polarity. We must be either negative or positive by choice or allowance. The traditional default polarity for humanity has been negative. Humans have understood that physical action creates experiences, and that many experiences occur unexpectedly for unknown reasons. We have believed that what we see is what's real. We have believed that we could only see things that already exist independently of us. According to quantum physics and spirituality, this is the opposite of reality. Consciousness is the origin and creator of everything. We are consciousness providing its conscious life force to our present awareness and creating our lives.

Consciousness is beyond thought and feeling. They are aspects of it, and our conscious and innate awareness envelopes as much of it as we allow. Consciousness itself is unlimited and infinite. It is everywhere and within every energy pattern and wave and being.

Within the empirical, human world, we have been given limits to our conscious awareness. They are prescribed by time, which exists because of the spin of sub-atomic waves/particles and everything that they are aligned with. Our limits have given us access to the dark side, the life-diminishing energetic realm. We have given our life force to these energies by aligning with them and giving them reality for us. We can transcend all of this, giving ourselves more wonderful lives and elevating all of humanity.

In our true Self, we do not need beliefs of any kind. To align ourselves, we can develop great sensitivity to our intuition. It is our connection with universal consciousness and our guide to Self-Realization. It has positive polarity, in alignment with the level of joy and compassion. This is the emotional level to feel for in every moment. It is present for us always in the quantum field, if we choose to be aware of it.

Exploring Our Sense of Being

Throughout our lives we have lived by the skills of our ego consciousness. It is who we have believed that we are. We've been unaware that we are much greater than we have imagined. We've had glimmers of deeper knowing and sometimes visions, inner voices and feelings of guidance in challenging situations, but we haven't known how to deepen this awareness and know it as a daily practice.

We designed our ego consciousness to operate within a limited spectrum of energies, so that we could experience the true feelings of these vibrations. The ego is not aware of the operations of higher consciousness. It functions according to its programming, and it fights for its existence, even though it expects ultimately to terminate. Meanwhile it fills our lives with personal dramas of all kinds, feeling victimized and separated from awareness of our greater Self.

How can we know if our ego is a limited expression of our unlimited conscious Self? We'll find out definitely when the ego dies, along with the body. Meanwhile we have the reports of many who have passed through the death experience or have otherwise learned to transcend the ego through conscious projection beyond the body. Beyond the ego, we never lose our presence of awareness as an individualized expression of universal consciousness.

Spiritual traditions have developed many techniques of quieting the ego mind and just being present in awareness. All require confronting and resolving our ego fears and self-imposed limitations. In a peaceful place, we can go within and seek emotions that truly turn us on with joy and ecstasy. We can imagine being aware in a realm of beauty, majesty and love. These are energetic patterns that we can align with. We may have resistance to this by our beliefs, telling us that it's not real, but when we do open ourselves in alignment with its energies, it becomes real in our experience. Our awareness expands to include a positive polarity magnetizing high-vibration experiences. We can allow ourselves to absorb it all and deeply enjoy our feelings.

In every moment, we can live in elevated vibratory frequencies in our awareness, and circumstances around us arrange themselves in alignment with our own energy signature. Fear disappears, because it's of a different polarity. Our entire lives become transformed into positive experiences. We radiate this energy throughout our body and our Being and into the energetic signature of humanity, opening up higher vibrations. This affects everyone in our presence and continues to feed back to us. Energetics work like this on every level and in every dimension. All of us can be masters of energetic alignments, making us masters of our lives.

Chapter 1. Significant Findings of the Quantum Sciences

Enlarging Our Intentional Awareness

Although we acceded to living in limited consciousness, we would never have intended to stay limited, because we are unlimited in our true Being. When we intend to expand our awareness into a higher dimension of living, we have the ability to create our energetic alignment with that realm, by being in gratitude, joy and compassion. This level of vibration of our state of being creates an alignment with the deeper aspects of our consciousness, being drawn into higher awareness of beauty, peace and kindness. These vibrations interact with the vibrations of our deeply-held limiting beliefs about ourselves. Our focus on an imaginary realm of beauty and love enables us to realize it into our reality.

When we begin to recognize and feel these energy patterns of positive, high-frequency emotions and thoughts, we can align with them. This choice notifies our subconscious innate self that resolution must occur in the contrast to our believed limitations. This contract is brought to our attention through our intuition and through events in our lives. In order to resolve this contrast in our consciousness, we can intentionally recognize the limiting beliefs for what they are. They were developed while we were in a state of negative polarity, resulting in an imprinting of energetic patterns that stimulate fear.

Through our perspective of compassion, we can understand how we have limited ourselves, and we can transform our polarized perspective into positive openness. We can anticipate wonderful experiences, regardless of whether we are facing chaos and threatening situations. By aligning ourselves with positive intentions to enhance life in our thoughts and feelings, we can experience the wonders of love and joy in our encounters. These vibratory patterns are always present in the quantum field that our true consciousness encompasses, and we can intentionally be aware of them through our imagination and emotions.

Because we are constantly creating our lives through how

we feel and think in every moment, we can make great changes in our lives by being aware of the quality of our perspective and intending to be positive and thankful. This creates experiences that stimulate these qualities in us, and it frees us from the bondage that negative polarity has imposed. Under the influence of positive polarity in life-enhancing energies, our stream of conscious life force expands, creating greater vitality and light in us. We can resolve our self-imposed limitations and expand our awareness into universal consciousness, resulting in our awareness of everything that we want to pay attention to, and inviting even greater experiences to come into our awareness.

Our Virtual Reality Presence

We are living in a virtual reality in which we own and control our own character. As we mature and travel around this world, we learn the rules of the game and the strategies involved in becoming better players. We're here to have fun and to learn about many aspects of this game that have important implications for our further development. We can learn to work with electromagnetic energies and their manifestations.

By being able to formulate scenarios that we create in our imagination with our emotions, we can control the forms and qualities of our experiences. We can create virtual experiences that we recognize, and that become actual experiences, when our beliefs will allow this. In virtual reality video games, new worlds have been created and in which your player participates in a certain spectrum of energetics. There are even paradise worlds.

In our empirical, virtual-reality world, we have complete freedom of choice in what we, as our character, can imagine and what vibratory level and polarity we want to experience, but we have been negatively polarized, as we participate in the hypnotic trance of humanity. In our game we can change our character and move ourselves into other dimensions. We are our only limitations to any kind of expression of life force. It is the constant vibratory stream of our thoughts and feelings that express themselves as our resonant energy signature with our own perspective.

Every perspective has limitations beyond which it will not align. Our focus of attention is filtered through our perspective, which consists of our beliefs about ourselves. If we intend to expand in conscious awareness, we can begin to understand beyond our beliefs. It becomes clear that we are a greater Being, whose consciousness is expressing our empirical presence as our character, our person. It is as if we just took off the headset and remembered who we are. This is the great realization of our presence here. After this, the game is really easy and can be great fun.

Opening to Greater Potential Realization

The more we align with the energy of the heart of our Being, the more intense everything becomes. Our life force gains freedom from limiting beliefs and can stream into our awareness with greater brilliance. We can realize aspects of consciousness that enhance a greater awareness beyond the physical. Our understanding deepens into greater compassion and generosity, and we can be aware of what level of vibration we are focusing upon. Everything we experience is a choice in polarity and frequency.

With an intense desire, we can experience the deepest pure love and ecstatic joy, whenever we want to focus on one of their manifestations that we recognize. While knowing what alignment feels like, we can choose the most heart-centered and inspired perspective in every moment. In alignment with our intuitive knowing, our ego consciousness can be an observer until called upon.

If we have not been relying on higher guidance in our lives, our ego consciousness does not know about it. In this case, we can break free by resolving our limiting beliefs. All we have to do is to raise our positive vibrations into a more wonderful spectrum of energy, until we are in a state of knowing our creative ability, how to use it and trusting ourselves to do the right thing. We have intuitive knowing to be aware of in every moment. This awareness requires intentional practice to acquire and maintain, until we just naturally have it.

Many members of our spiritual family are on this planet currently. We are awakening to our expanded awareness and higher level of joy. If we have difficulty being positive, but we want to be, we can do so by finding a deep breathing technique that takes us beyond empirical limitations, such as controlled hyperventilation, or we can just laugh uncontrollably for 10 or 15 minutes. These practices can take us into a positive state of being, in which our intuition becomes more recognizable, and our ego is mystified.

When we are most receptive, we can align ourselves with our intuition, which flows into us with our life force from the consciousness of the Creator. We are guided by the intelligence that controls universal consciousness. As far as we can imagine and beyond, we have the ability to expand our awareness into universal consciousness. Versus the historical 7.83 cycles per second, it appears that the rising resonant frequency of Gaia is in the 40-100 cycles per second range. This new level is a good spectrum to align with. It's filled with wonderful energetics of purity, abundance, freedom and joy, and it interpenetrates us right now.

Expanding Our Personal Awareness

We are designed to be however we want ourselves to be. We have the conscious design of freedom in every aspect of our Being. This is an aspect of the consciousness of the Creator of all.

Since we are fractals of the Creator, we have the same infinite abilities throughout our psyche. They are waiting for our recognition and invitation for them to manifest for us. It is as if each of us is a rain drop arising from the ocean, forming clouds, and eventually falling back into the sea. It is all water, just as we are all the consciousness of the Creator, appearing in myriads of forms and emotions.

In our experiences as humans, we have not believed that we are the Creator of all. In our true Self, we are the personal consciousness of the Creator. When we choose to align with this awareness, we become the masters of this dimension. Our experiences arrange themselves in resonance with our state of being. To our ego-conscious self, it appears that we live a miraculous life, but it's all about our polarity and vibratory status. This is how our consciousness expresses itself as the energetic patterns that come into our experience.

We control the quality of the energetics that manifest for us by the polarity and vibratory patterns of our predominant thoughts and feelings. It is how we imagine ourselves to be in every moment. We have been trained to imagine ourselves as separate individuals with an awareness limited to the empirical world and to our limited mental and emotional capabilities. Our ego consciousness cannot know universal consciousness and infinite Being, but it can accept higher guidance, when we become aware of our true intuition.

Once we resolve our limiting beliefs, and no longer recognize them as real, our awareness is naturally unlimited. By choosing to be in a prolonged state of positive high vibrations of joy in our imagination and feelings, we elevate our personal energy signature. We can rise above polarity and frequency into eternal Oneness. This attracts beauty, freedom and abundance into our experiences and radiates the love and compassion of our heart all around us. It elevates the energy signature of humanity and aligns with the rising resonance of Gaia.

Love Song of Our Heart

The most important aspects of our lives cannot be seen or touched, they can only be felt in the heart of our Being. We have been accustomed to receiving direction from outside ourselves, but we can know truth only in our own intuitive knowing. We are individualized expressions of the love and joy of our Creator, able to interface consciously with every entity in our experience.

From quantum physics experiments we know that everything in our world is conscious, down to the tiniest sub-atomic particles/waves comprising everything. All consciousness participates in the unconditional love and life-enhancing energy of the Creator. Without these qualities, nothing could exist. Our own degree of consciousness is as great as we are open to realizing, depending on the intensity of love that we are willing to align with and express.

In a state of serenity, we can ask our greater Self to draw our awareness into the light and joy of our true Being. With the intention to know and be our divine Self, and the willingness to practice, we can penetrate our limiting beliefs about ourselves, and expand our awareness into universal consciousness. In this state we can sense our intuition as our inner knowing. We can feel the quality of the life force that constantly streams into our consciousness from the essence of our Creator, endowing us with Creator consciousness.

Within our stream of life force flows unconditional love for all conscious beings. This level of vibration includes all of the higher feelings of joy, compassion, freedom, sovereignty, and beauty. It also includes realization of our eternal present awareness. We may hear inspiring inner music. This is the song of our heart. It is the essence of our Creator flowing through us, expressing Itself through our free will and our attention.

Resolving Doubt about Our Creative Ability

If we feel that we are dependent upon factors outside of our own being for our sustenance and enjoyment of life, we do not believe that we have the ability to create everything we want. Why do we have this doubt? We have kept our consciousness compartmentalized within the boundaries of the empirical world. Our understanding has been limited to our experiences in the third density of human reality, where there is misunderstanding of the creative process.

Because we are infinitely powerful creators, we have had to drastically limit ourselves to play the part of humans on this planet. This is why we compartmentalized our consciousness; otherwise, our lives here would not give us the depth of experience we wanted in the realm of negative polarity and fear. We have expanded universal consciousness through our experiences in Earth life.

In our true Being, there is no polarity between positive and negative. We are pure presence of awareness with infinite abilities to express ourselves, and we participate in the universal consciousness of the Creator. In our human form, we create with our thoughts and feelings and everything that enters our awareness, when we align ourselves with its polarity and vibratory level. It doesn't matter if we are for or against something. What matters is our engagement and energetic alignment.

What we desire already exists for us in the quantum field. We just need to align with its energetic patterns in order to recognize it as real and be able to experience it. When we can feel ourselves living in the kind of experience we desire, we align ourselves with its polarity and energetic frequency patterns. With our energetic alignment, we can recognize and accept the experience as it become real for us. If we're motivated to practice this, and we have the intention without doubt, we can accomplish living a wonderful life and help elevate the vibrations of others.

Because we are creators through our ability to modulate energetic patterns with our thoughts and emotions, our doubts express negative energies, which manifest in the human dimension as failure of creativity. We have not known that we are creators with our energetic alignments. This is what we do as humans, we expand the experiences within universal consciousness in ways that would not be possible for the consciousness of the Creator, which we participate in.

Transforming Doubt into Trust

The essence of our Being is our awareness. It is beyond time and space and even beyond our imagination. It is eternal presence of Self-Awareness. When we are pure self-awareness, we can be aware of all of the energetic patterns that we envelope in our consciousness. We can focus on any of them and bring them into alignment with us through our power of energetic modulation with our thoughts and emotions. To be effective with our abilities, we can be confident in our eternal Being beyond the body. This we can know intuitively in the energy of our heart.

In this perspective, we can recognize that our current human experience is a projection of human consciousness. We all focus our awareness on the spectrum of energy of this dimension of existence. With our continuing focus, we maintain its existence. There are other dimensions that we can evolve into, and if we develop a strong intention to focus on that spectrum of energy, it can become real in our experience.

Developing trust in what we intuitively know is important. For this, we can engage in meditation or any method of allowing ourselves to drop all concerns and plans and just be present awareness. Once we can just be present awareness, we can move beyond the body consciousness to include the etheric realm, which is imperceptible to most humans. From within its presence, we can be aware of greater light and beauty.

Intuitively we can know that, as humans, our consciousness is bound by self-imposed limits. These are our limiting beliefs that are based on fear and doubt. If we examine these energetically, we find that they are negatively polarized. By transforming the polarity to positive, we can resolve them. We can do this by intentionally focusing on alignment with the heart of our Being, the place of deepest love and greatest joy. This is our natural state of Being, when we have moved beyond fear and doubt.

To know confidently that we always have everything we need, as abundantly as we desire, and as long as we remain positive, transforms our lives in wonderful ways. We gain confidence as we learn to rely on our intuition for guidance in every moment through our emotions and symbolic encounters.

Any negative patterns in our environment can pass through our energetic field, as if we don't exist. It's a matter of dimensional veils. In our world of duality, we live with both positive and negative energies, and we can learn to deal with the negative ones by transforming them or letting them dissolve out of our experience into another dimension. This is how electromagnetic energy patterns interact. If there's no alignment in resonance, there's no encounter.

Aligning with Life-Enhancing Energies

Quantum physicists have shown that there is universal consciousness, and that consciousness is the source of everything that we experience. The empirical world is an expression of the consciousness of humanity. Through the focus of our attention and alignment with the empirical spectrum of energy, we recognize it as real for us. It is the recognition and alignment of all of us that creates its manifestation in our collective experience. Our collective conscious life force creates and maintains our reality according to our predominant state of being, our vibratory level of awareness and polarity.

We have been trained to live with a certain level of stress and fear, and to believe that we are mortal. Outside of our own consciousness, nothing compels us to believe that we are mortal and subject to any negative energies. We have had false beliefs that have kept us enslaved in a perspective of negative polarity. Our vibratory polarity and frequency patterns attract compatible energetics to come into our experience. How we react to them on an energetic level determines the quality of our experience.

If we can react to ego-threatening situations, while being in a state of compassion, gratitude and love consistently, the circumstances of our lives become situations that stimulate those vibrations in us. Every moment can bring change of any magnitude, depending upon our state of being. From living in duality, we have created a knowing of life-diminishing and destructive energy patterns within universal consciousness, and we may have been locked into reincarnations here for eons; however, we can open ourselves to our true Being, whenever we desire to realize our eternal presence of awareness.

When we are in secluded, beautiful and majestic places in nature, and we can just be present in awareness, we naturally drift into positive polarity and high vibrations in alignment with the resonant energy of Gaia, Spirit of the Earth. The more we can be in this state of being, the more our lives are transformed into life-enhancing experiences of freedom, abundance and sovereignty. As our awareness opens to higher vibrations, we begin to transcend the compartment of consciousness that is the world of separate human experience.

There is nothing outside of our own consciousness. Our true awareness is unlimited, as are our creative abilities. Once we realize our eternal Self, and we have become mentally and emotionally clear, we become masters of the empirical world. We know how energetics work, and we can transform any energy patterns into positive, high-frequencies in alignment with us. These are the energetics of the universal consciousness of the

Creator, and we can align ourselves with them through our intention and desire, and our eagerness to practice.

Analyzing Our State of Being

In our true Being, we live in our consciousness and have our presence in the universal consciousness of the Creator. Everything that is natural for us arises directly from universal consciousness. When we are positive and coming from our heart energy, we can feel the vibratory level of our life force. We can know the guidance coming through our intuition, and we can follow through in our actions in ways that align with our inner guidance. Aligning with our intuition may require practice in developing sensitivity in a positive, life-enhancing perspective.

Our role in life is to create new energetic patterns in the expression of life. We are expanding consciousness through our experiences. We've experienced a vast spectrum of duality, especially the dark, negative energetics, as we participated in a clever compartmentalization of consciousness to experience energetic patterns that would not have been possible within universal consciousness. When we decide to return to full consciousness, the path is guided by our intuition, which works through both our imagination and our emotions.

If we desire to know what is most natural for us, we'll be aligning with our intuition. To be truly receptive to the energy of the heart of our Being, we must resolve all of our limiting beliefs, so that we can be open mentally and emotionally. Only our self-imposed limitations can separate us from our natural perspective of eternal, infinitely powerful creative presence of awareness in alignment with life-enhancement. We are naturally loving, joyful, compassionate and grateful. These are the feelings that accompany a positive perspective and lead to the vibratory level of Self-Realization.

Our creative abilities know no limits. Once we are in resonance with our eternal Self, we gain our own self-trust and stop interfering with our own creations, allowing them to manifest in our experience. For us to exercise unlimited creativity, we must have clarity and presence of awareness. This establishes a state of being that radiates the quality of vibrations for our creations.

Our presence in universal consciousness can transform our lives among humanity, because we become exempt from all negative energy. We do not align with it through engagement, and it disappears from our experience into another dimension. If we so desire, we can enable ourselves to live in eternal awareness of love and joy in the universal consciousness of the Creator, while living in a beautiful environment on Earth.

Empowering Our Lives

In this lifetime, we have come to settle our issues and resolve our personal dramas. We are our own heroes and villains, depending on our chosen polarity, positive or negative, love or fear as its basic energy. The quality of our experiences depends upon our own vibratory level. Regardless of what our intentions might be, we align ourselves with a polarity and vibratory level. Anger is anger, regardless of whether it's righteous or diabolical. When we oppose or resist something, we align with the same quality of energy as the force that we resist.

This may be difficult to understand, but demonstrating against war is the same energy as supporting the war. The way to resolve this is from a positive perspective of compassion and understanding. What matters here is the polarity and vibratory level of the energy that we align with and radiate into the quantum field. This is what manifests experiences for us and influences the energy resonance of humanity.

Even when we know that we've been programmed to believe in our limitations, we find that they go deeply into our psyche and do not want to be recognized for what they are. We've allowed our consciousness to be tied down and bound within a limited compartment of awareness by limiting beliefs about ourselves.

Since we know from quantum physics that there is a universal consciousness, everything that exists participates in this consciousness. Consciousness is the creator of everything, and the vibratory level of creative energy is life-enhancing in the most beautiful and wonderful ways. This is the quality of energy that envelopes us. Out of universal consciousness, we arise as personal Self-aware Beings. Just as universal consciousness is eternal, so we exist beyond time and space in our conscious Being.

In our human form, we have our connection with universal consciousness through our intuition. It comes to us constantly in the conscious life force that enlivens us, but it is not intrusive. In our limiting perspective, our ego consciousness, we have cut our awareness off from our intuition. To expand our awareness to our higher guidance requires our intention. We can imagine ourselves being our higher Self and enjoying our unlimited abilities based in joy, gratitude and love. On this level of vibration, we can be sensitive to our inner prompting. We can be aware of everything in our lives from a perspective of intending to expand our consciousness into a higher dimension of living, and we can know deep within everything we may want to know in every moment.

2.

Resolving our Limitations

∞

Unbinding Our Limiting Beliefs

Whether we realize it or not, we are the designers, directors and producers of our own lives, regardless of our apparent situation. We experience our current lives because of how we have created them, probably unintentionally, in the past, not necessarily the forms, but definitely the vibratory quality. The limits that we live within are self-imposed and resolvable. If we have a clear intent, we can resolve the limiting beliefs that are hidden deeply within our innate consciousness by examining them in the light of what we know in our heart is intuitively true.

We have learned to regard the empirical world as solid and available for the stimulation of our senses. This is a play within our own consciousness. Quantum physicists have found that what we recognize as matter is actually electromagnetic energy patterns. Even what we recognize as sub-atomic particles, which comprise everything in the empirical world, are actually patterns of conscious, spinning energy waves, which stimulate recognition through our senses. It is our conscious recognition that gives them the quality that we recognize as substance. We can change our recognition by changing our perspective. We can intentionally unlimit ourselves, resulting in a different world of experience.

When we realize that the empirical world exists as we believe it to be, and that it is our beliefs that keep us locked into experiences within a limited band of wave patterns, we can decide to open ourselves to awareness beyond time and space, while modulating the energy patterns in our experience. Our emotional and mental state of being has an energetic vibration. This is what magnetically attracts energy patterns in the quantum field that align with our vibrations. We may go from positive to negative polarities and back again. We may imagine having unscrupulous rulers, and everyone around us may imagine the same thing. This need not affect us.

We are free to have our own perspective. We live within the vibratory patterns that we entertain in our awareness, those that we dwell upon and align with. When we choose to imagine feeling joyful and unlimited in every moment, we enter this state of being, and we no longer interact with negative, low-vibration energies. We can live in a higher dimension of harmonics and light by being in nature. We can feel ourselves aligning with the energies all around us, and we can open our awareness to universal consciousness and unlimited creativity.

Our Current Timeline

If we have resolved our limiting beliefs about ourselves, we are free and can have unlimited creative abilities. As long as we maintain our thoughts and feelings in the spectrum of love, we become exempt from the low-frequency energy of the human drama. Our inner guidance knows how to interact always from the perspective of the unlimited consciousness of the Creator.

We can intend to be our true Self. Through intention, we can realize our expansive consciousness as we align ourselves with the flowing energy of our life force and our conscious personal awareness in the vibration of unconditional love. Our intention to be loving beings can lead us into alignment with the consciousness of the Creator. In this level of vibration, our personalized self, our ego, can relax, observe and enjoy flowing in higher guidance. The ego knows that its guidance for us has been limited. We can thank this aspect of ourselves for taking us through all manner of fear and deeply-held dread of termination. We now know the feelings of all of these vibrations. In our expanded Self, we now know the true meaning of compassion in our deepest feelings. We can know that we have designed our human experience to gain emotional depth in our expanding consciousness. The low-vibration human experience has been only for the bravest of souls, because it's a great challenge to be able to raise our vibrations in true commitment, trusting the feelings and intuition of the heart of our Being. We've been taught to ignore these feelings completely. They're still present, if we want to experience them. They require our attention and recognition in the vibration of gratitude.

Once we begin to imagine our personal Being as eternal, we can become unlimited in our awareness in all dimensions. Our Earth-human experience has served us well, and we can now rise to the challenge of transforming ourselves out of the vibrations of fear, by aligning with our higher inner guidance. Our conscience is part of this guidance. It also requires our attention and thankful recognition. The more we can have this awareness, the more prominent our guidance becomes.

As we align more with the energy of our heart, we naturally exist in high vibrations and attract life experiences that have the same alignment, regardless of our current life. We become more aware of compassion in our interactions and more loving with one another. It can be a mutual expansion of our conscious awareness. It is in alignment with the rising vibratory resonance of the Spirit of the Earth.

The compartmentalized, fear-based consciousness of humanity is fading. We are on the path to expanding consciousness in alignment with the Earth and all conscious beings here. The light is growing in every heart.

Resolving Our Barriers to Higher Consciousness

We have been accustomed to thinking of ourselves as subject to our circumstances and to what others do to us. We have recognized through our senses, that reality exists outside of ourselves, and that other people, animals and things are different beings from us. We have learned that life happens to us, and that we must do our best to survive and thrive by the intelligence and skill of our ego consciousness.

If we feel a lack of love or abundance or joy in our lives, it is always because we are self-limiting the flow of life force that we are designed to enjoy. We may not know how we are causing ourselves to suffer lack in anything, but we can open ourselves to fulfillment whenever we are ready. We do not need help from anyone else. This is a personal transformation that must occur for each of us. It is a matter of remembering and realizing the truth of who we are in our most expanded essence beyond time and space.

Quantum sciences have determined that everything that exists is conscious. Everything is composed of sub-atomic particles/waves that display universal consciousness. Water molecules have shown a wide range of conscious responses to human actions and thoughts. All plants display emotions and vitality depending upon the energetic frequencies in their environment. Animals are the most vibrant in their feelings and interactions. We humans have been the most destructive and thoughtless in our interactions, because we've been focused on experiencing low-vibratory energy patterns in our mutually-created, self-limiting environment of the empirical world. We have struggled for survival apart from our higher guidance, which we have ignored.

We are so much more than human beings. We are constantly being created and enlivened in the essence of the consciousness of the universal Creator. We are the masters of life, constantly creating the qualities of our experiences according to our personal beliefs and preferences. We can learn how to focus our attention intentionally on the quality of energetic patterns that feel best to us. We can intentionally cut off our attractions and connections with destructive energetic patterns and replace them with high-frequency visions and feelings of personal fulfillment. Instead of living by hollow intent, we have the ability to create wonderful and fulfilling lives through our recognition and feeling for greater awareness of love and vitality.

The Evolution of Human Experience

As we begin to awaken from eons of living under oppression and fear, not knowing how to free ourselves, the light is dawning in our awareness. In this lifetime we have a destiny to open ourselves to higher aspects of Being and to elevate ourselves above the low vibrations that we have allowed to enrapture and enslave us. We can now recognize that we have never been victims or oppressors without our cooperation. We have aligned our energetic presence to the vibratory level of the low-vibrating energy patterns that we chose to focus upon. We have done this by feeling anger, resentment, hatred, judgmentalism, jealousy, starvation, suffering and all other forms of fear. Yet, we have always had the choice of how we want to feel in any situation, regardless of what we experience reflected back to us from our former creations of our perceived enemies.

By our focused intent to be only compassionate and loving

toward ourselves and all conscious beings, we can be more powerful than any negative beings by being positive and rising above their spectrum of frequency and becoming invisible to them. This requires constant alertness and training, until we have no presence in the lower vibrations.

We're all being challenged by the unmitigated proliferation of low-frequency, destructive, life-taking energy of those who refuse to acknowledge the life force of unconditional love constantly flowing into us from the universal consciousness of the Creator of all. They need our life force to survive, since they've closed themselves off from their own source of Being, and they feed on us through our low-vibration energy. They need us to be fearful, angry and depressed, and to align with their frequency by fighting against them or imagining our defeat. They need us to engage with them.

We can recognize that the entire human experience is played out within a compartmentalized portion of our innate consciousness. Through deep meditation, being open and inviting experiences in our consciousness beyond the body, we can realize our eternal presence of Being. It is our Self-Realization that frees us from the realm of fear and opens us to know and experience the fullness of life in every way.

Our Stages of Awakening

On this planet humans have lived in survival mode for eons. We've been subjected to low-frequency energies that have kept us in fear. This is an unnatural state for us, and we could not experience it without our permission. We have willed ourselves into our current situation. We have opened ourselves to experience the effects of low-vibration energy. We have lived in sympathy with victimhood on a deep level. People are now voluntarily being injected with trans-human technologies designed to result in planet-wide permanent servitude with no choice of escape.

Until now, this has been a choice. It is not a necessity. We can change our experience whenever we choose, once we realize it and begin to look for a way to transform our ourselves.

Once we realize that we do not need to subject ourselves to pain and suffering, we can find an increasing amount of guidance available for us. We will discover that we have created the quality of our own experience, which we have intentionally shared with humanity. It is through our predominant focus of attention and what we feel, that we create an energy signature that vibrates with the electromagnetic polarity and frequencies of our thoughts and emotions, with an amplitude of our intensity. This is how we inhabit our own consciousness, attracting encounters and situations that vibrate in alignment with our energy signature.

By consistently maintaining a perspective in the love vibration, we can recognize that we can create any quality of experience we want by feeling and imagining encounters and scenarios of high vibrations, and by accepting the quality of every encounter with compassionate wisdom, gratitude, forgiveness and love. These are all natural expressions of the energy that carries our conscious life force at the heart of our Being. By penetrating the depths of our awareness, searching for the source of our inner light and self-consciousness, our awareness begins expanding greatly with our understanding. We enter a higher dimension of consciousness, from which we understand everything about our empirical Earth human experiences.

As we begin to enter a present state of pure awareness, our abilities become clearer, and we can feel the energy of our heart guiding us through our intuition. The circumstances of our physical experiences arrange themselves to align with our own rising vibrational signature. It may seem like empirical magic, but it's just the interaction of energy patterns of the quantum field, enveloping us, attracting and repelling one another. We are the directors of our own energetic expressions and attractions.

Our True Personal Power

Innately we all know the eternal Being of ourselves. As humans, we've compartmentalized our awareness to be unaware of our true Being. We live as if we are the ego-consciousness that we use in our compartmentalized experiences. We've created our persona in ways that would attract certain kinds of experiences that would provide more depth in our understanding of energy patterns and life.

We created our own deep-seated beliefs to keep our awareness within the compartmentalized human energy resonance. We can resolve all of our limiting beliefs by intentionally striving to be our Creator Selves and then being Ourselves, our pure Being in expanded awareness of unconditional love. In this dimension of vibrations, we can become fully conscious beyond the disappearing limits of our beliefs. We can move beyond ego-consciousness to universal consciousness in the Being of the Creator. This is the essence and source of who we are and what we are capable of.

Who we are is our conscious awareness without our bodies. We are pure awareness with unlimited abilities. We are clear and transparent in high-frequency resonance. As we begin to recognize our true Being, our conscious awareness expands, enabling us to realize the essence of our Creator. We can know from within, that every conscious being shares the essence of ourselves. The same life force enables and enlivens all that exists. It is our conscious life force constantly streaming through us in unconditional love.

We receive our life force within the limits that we impose upon ourselves. Any beliefs that are not life-enhancing must be resolved, because they keep our vibrations aligned with their energy. This is the source of doubt and fear. Once we recognize that we are eternal awareness, we can know that we are the creators of everything we experience. Doubt and fear pass away.

We can be aware of feeling high-vibration emotions as much as possible.

We can interact in every encounter with compassionate wisdom, acceptance, forgiveness, gratitude and understanding. All challenges are messages requiring this approach. Once we can stay in the love vibration, we continuously create high-quality experiences, because we are designed to create what we recognize. The energy patterns that we envision and recognize become material for us. This is the miracle observation that launched quantum physics. Physicists found that when they used instruments to observe photons, the photons changed instantly, from being quanta of energy into visible light in our empirical spectrum. This shows that recognition of energy patterns and alignment with them on our part causes material manifestation. This is how we are the creators.

Resolving Our Issues

The human resonant frequency is a limitation for us, but we can reach for the outer edge of human consciousness, if we intend to raise our frequency. We can achieve complete clarity of mind and emotion by paying attention to what arises in us when we're challenged or threatened. If we examine the belief that comes up, we can resolve it, through acceptance, gratitude, forgiveness and love.

We are all actors in a drama, playing roles that we've agreed to play. The entire scene is a complex pattern of electromagnetic waves that our consciousness interprets as our empirical reality. Determined by the vibrations of our personal energy signature, the qualities of our experiences manifest our energetic presence, vibrating at the level of our predominant thoughts and feelings. The qualities of everything we create are determined by our emotional vibratory pattern and our polarity between life-enhancing and life-diminishing energies.

We have become emotionally attached to our accustomed way of life and all of the people we feel close to. This is a magnetic attraction that holds us at the conscious level of resonance of those we're attached to. As long as we feel attached, we will not change our awareness. The qualities of everything we create are modulated by our emotional vibrations. This is our magnetic alignment ability, attracting resonant patterns of experiences.

When we are clear, we can truly love without attachment. When we're ready to detach ourselves completely from everyone and everything that we have been attached to, we can make the leap deeply into the positive polarity, high-vibration level of love. In this state of being, we are aware of our intuition in a very sensitive way, and we are guided in every moment. We do not need to make plans, because we always know what to do in every moment. We become able to play our roles by listening to our inner teleprompter. We are active participants in the human drama, and we can change scenes and actors at will.

By resolving all of our limiting beliefs, based in life-diminishing energies, we can open ourselves to unlimited freedom of awareness, moving into universal consciousness and our identification with the Being of the Creator, out of whom everyone and everything constantly arise.

Bringing Our Innate Abilities into Action

We are advanced Beings, and have been our own awareness before this incarnation. As children, we played with our unusual abilities, learning to use them; although some of us turned away from them, under pressure from others. Through our interactions with others, we created our personalities, and we've kept modifying them ever since, trying to align with the vibratory spectrum of humanity. The roles that we're playing can now be elevated by our intentional thoughts and feelings, as we participate in the rising resonant frequencies of the Spirit of the Earth.

We can awaken our abilities and wield them for the good of all conscious beings. Once we realize that our awareness is eternal, we can become masters of our lives. We no longer need our accustomed personality. We can be incredibly free to be our deeper Self, dependent on no one, because we can be guided by our intuition, in which we can be aware within the universal consciousness of the Creator and the energies of the unified quantum field. We can create at will, operating in alignment with the resonance of the heart of our Being.

If we can attune our feelings to the high vibrations of love and serenity, we become unavailable to the low vibrations of fear and dis-ease. When we drop our old personal beliefs and ways of being, we can open ourselves to the freedom and sovereignty of our true Self. Our presence becomes brighter and more radiant. Our imaginary and visionary abilities, in alignment with our emotions, can be the creators of a wonderful life of high-vibration experiences.

We express ourselves with electromagnetic wave patterns emanating from the heart of our Being. This energy is filtered through the spectrum of our personal beliefs, to vibrate as we imagine and feel. Our mental processes operate on electricity, and our emotions operate on magnetism. Only when the two intersect do we perceive an electromagnetic wave pattern. How we modulate it with our thoughts and feelings becomes the quality of our experience.

Personal beliefs are the greatest limitations to our total awareness. By resolving all of these, we can find our path to the higher-dimensional, unconditional love of our true Being. This is where we can feel so much joy and gratitude.

Unlimiting Our Conscious Awareness

In uniting with our higher Selves, we can feel the most wonderful and deepest love, more than we can imagine. It is beyond time and space in the enveloping quantum field as energy patterns that we can recognize. When we do, they transform our awareness beyond the empirical world. Our emotional nature can help guide us into higher vibrations of thinking and living. By our ability to focus our attention where we choose, we can bring our emotions into more wonderful states of feeling through intending to be grateful and joyful. These are the vibrations of the infinite Creator consciousness. They stimulate genius abilities beyond the physical and can be brought into manifestation in our physical experiences by our recognition of their reality.

To recognize higher-vibration energy patterns, we must clear ourselves of personal interests based in fear or its potential. We can know that we are our eternal present awareness with unlimited creative abilities. We can know everything about what we focus on, when we are aligned with our greater Self, because we have access to universal consciousness. This is attainable for all of us, because it is who we truly are, and we can know this through our intuition.

We are awakening to the possibility that we are more than our empirical body presence in the compartmentalized consciousness of humanity. We are more than our beliefs will allow. If we can resolve them all, we can attain clarity of Being, because it is only our intentions and limiting beliefs that keep us within the limited conscious energy spectrum of humanity. All of our beliefs were created from fear. They have a negative polarity. These energy patterns were created and remain in place by the recognition of all of us, but we don't have to remain in this dimension exclusively.

We can expand our awareness greatly into higher dimensions of greater clarity and brilliance. We can enter the vibratory spectrum of gratitude and compassion. In this vibratory resonance we can experience life on this planet from a perspective of deepest enveloping love and joy. Every situation that we face is a communication between our ego-conscious self and our

Creator essence, which is our inner guidance. The ego self needs our deepest gratitude, love and assurance of being. It is part of our personal Self-awareness, and it can now relax and observe. Its role is complete, and we have learned what we needed to develop our wisdom and compassion.

Intentional Living for Conscious Expansion

We can improve our lives by wanting and constantly intending to feel more loving and compassionate. We can also constantly intend to be aware of our conscious presence and the intuitive knowing within the heart of our Being. Our intentional focus on higher vibratory feelings and thoughts draws the same quality of experiences to us. We have complete control of our vibratory level at all times, and we constantly choose the vibratory patterns that we focus our attention on. Our energy levels express themselves as our personal energy signature. The radiance from our energy signature attracts energetic patterns that resonate in alignment with us. These appear to us in the quality of our experiences.

Our true abilities are vast, far greater than any super hero. To access them, we need to clear ourselves of all of the beliefs that we designed to keep us from realizing our true Self. We did this, so that we could fully understand the human experience, with its low-vibrational attractions and threats that we believe are real. We give them our life force by aligning with their negatively-polarized vibratory level.

As we awaken from our personal reality play, we can begin to realize our expanding present awareness. We must intentionally shift polarity from negative to positive. This is a leap in consciousness from the realm of fear to that of love. Our ego cannot conceive of this, because it has no higher guidance. We designed it this way. If we choose the way of love, we are shown our eternal present awareness beyond time and space. We can intend to be always just present in awareness, while being clearly attuned to our intuition, which is our connection with our true Self. This level of vibration elicits ecstasy on our part, it is so wonderful!

By constantly intending to open ourselves to the presence of our awareness in every moment, we gain an elevated new perspective, and we can move beyond ego consciousness through great love. Regardless of what we may be encountering in the world of humanity, the most important state of our being is to be in a spectrum of high vibration thoughts and feelings. When we are living intentionally with a focus on high-vibratory energy patterns, we can imagine and feel as if we are participating in scenarios that are wonderful and exciting. Memories and expectations can disappear. There is only the present moment in all of its richness that allows for our creative energetic alignment, constantly and forever. This is what our energy signature radiates into the quantum field for manifestation into our experience by attracting compatible energy patterns in whatever dimension we're vibrating at.

Evolving Beyond Our Beliefs and Fears

The function and basis of our beliefs and attachments are available for us to know. With this knowledge, we can expand our consciousness, because we then can allow ourselves to. We have self-imposed limitations and attachments that we align our vibratory patterns to resonate with. As long as we are bound by our alignment with those vibrations, we cannot unlimit our conscious awareness.

We created our beliefs and set them deeply into our subconscious in order to deprive ourselves of universal consciousness. We needed this for our human experience of living without higher guidance. We had to have no idea of who we are in our complete Being. As a result, we have learned what life is like in the human spectrum of vibrations. When we are ready to expand, we can intend to be aware of our limiting beliefs and preferences.

Shifting from fear to love, from negative to positive, is a leap in consciousness, because without higher guidance from our intuition, we cannot know what would happen to us. We can recognize our alignment with the qualities of vibrations of our attachments. They are all negatively polarized, because we fear that we may not always have what we most want. If we were to change polarity to a positive perspective of a compassionate and loving state of Being, we would not yet know that we can always have everything we need to survive and feel fulfilled.

We can feel gratitude for our opportunity to experience the heights and depths of human life, the passions, hopes and fears. We can forgive ourselves for wanting to know what suffering is like. We can love the entire experience and all of the characters, good and evil, that have taught us much deeper compassion and love than we could have imagined, as well as the sensuous pleasures. It is all a creation of human consciousness.

By being sensitive to the vibrations of the conscious life force that we receive in every moment, we can recognize the qualities of all of the energetic expressions that we confront. We can intend to be open to the presence of higher-frequency beings, with whom we can align through the energy of our heart. This draws us into the positive polarity of love. Our entire being can be positively polarized and vibrate with the enhanced vitality of a higher dimension of awareness. In the high-frequency, positive polarity of love, we can realize that we always have what we need to be comfortable and to enjoy life. This is a higher vibratory quality of living.

We can always know intuitively everything we want to know. Practicing to be sensitive to our intuition is the beginning of all knowing. Our intuitive prompting and guiding is always present for us to be aware of. It is not intrusive, and if it is disregarded for years, it becomes very faint, but never disappears.

Continuing to Evolve

For anyone who may be on a spiritual path, but the journey is not yet clear, pay attention to how you feel in any moment, and then recognize the quality of feeling in the conscious beings around you. All of them, plants, animals, Gaia, even the Sun and stars. These are all conscious beings, who are expressing themselves as their chosen forms, as are we. We have planned our entire human experience, so that we could develop deep compassion and greater wisdom. We had to explore the depths of low-frequency vibrations, so that we could develop our emotional awareness and intentionally choose greater alignment with vibrations that feel the best that we can allow ourselves to realize.

If we look for them, we can recognize our limitations. Our essential Self knows all of them, and allows our personal drama to play out. We created our ego consciousness, with our self-imposed false beliefs and limitations, to learn to choose more elevating states of Being. We can have emotional and mental alignment with the vibratory frequency of the energy patterns that we focus our attention on. By maintaining the emotional vibratory level of joy as much as possible, we can transform our false beliefs and resolve our limitations. We can be aware of our eternal presence of Being. Apart from our body and the empirical world, our consciousness is universal. We arise out of the consciousness of the Creator-Being as fractals of Self. We are expressions of universal consciousness and can know the entirety of it. We are not mere humans. This is only our current expression, which has largely served its purpose for us.

We have the opportunity to step through the portal into a higher dimension of Being. The vibratory patterns here are of positive-polarity, high-vibration feelings and thoughts. This is the realm of truth in everything. The energies around us are transforming from negative to positive polarity, from fear to joy and are rising in frequency. This expanding consciousness is felt in all conscious beings. We are being drawn to awareness of entering the universal consciousness of the Creator and recognizing our awareness beyond time and space. We can be aware of our eternal presence of awareness with infinite creative abilities.

Our ego consciousness cannot grasp any of this. We created it to struggle to get us through the human experience. It had to operate without intrusive higher guidance, and preferably be unaware of it. So here we are, limitations and all. Now we have the opportunity to transcend our ego consciousness joyously. We can be guided in every moment by our intuition, through the conscious life force flowing into the heart of our Being. Once we place our attention on our intuition and can recognize it, we can have a productive inner dialogue for higher guidance.

Developing Cosmic Consciousness

We are unlimited Beings of conscious awareness. We can manifest as persons in any dimension, but the persons are each only one expression of our Self. Our Self has unlimited consciousness, and personhood is one of many of our possible expressions. Our persons living as humans are one of our expressions. As fractals of the Creator consciousness, we are free to think and feel however we want.

Our Self has given us a compartment of energy patterns to experience and create within. It is the empirical world. We have all been left with our own boundaries and preferences in order to have the kind of experiences that guide us and challenge us to evolve multi-dimensionally. We must learn how to resolve all negative-polarity, low-vibrational frequencies abiding within. This may be a difficult endeavor. It is where we wrestle with our angelic Self and is what causes us to suffer and to begin to close ourselves off from our life force.

We maintain a certain darkness hidden within. We need to become aware of these low-vibration energy patterns that we're hanging onto. Once we can recognize them, we can realize that we are able to change the resonance of these false perspectives through being compassionate and loving toward the energetic patterns of experiences that we are facing.

We can feel and imagine ourselves as infinitely powerful Creators. Just feeling as if we are in this state of awareness is a powerful experience. If we do this often enough, we become able to believe in its reality, and it becomes real for us. This is the direction of the rising resonant frequencies of the Earth and of humanity. We are learning what is true in all aspects of life. Human consciousness is expanding into a higher dimension of energetic patterns. It is a realm of positive polarity, high-frequency living of love and joy. We can feel really good in every moment.

We can become masters of life on this planet and in other dimensions. Since everything that we experience is patterns of electromagnetic waves, we can open our awareness to a larger spectrum of energetic patterns. Our awareness is limited by our own beliefs and intentions. These we can examine for their quality of energy. We can then resolve them from a perspective of compassionate wisdom. This is how we can free ourselves from the boundaries of human consciousness and enter a reality of wondrous joy and goodness.

Interacting within Humanity's Energetics

There is intentional genocide being forced upon the people world-wide. The key persons that can control it all are the military leaders, including the CIA, but they depend upon the bankers for money. So, the bankers buy protection, and both groups control nearly all of humanity for profit and entertainment. It is the rise of fascism everywhere. This is the ultimate move by the lowest-vibratory, most destructive energetics of the negative elite, but they can be defeated by the energetics of positive polarity, high-vibration patterns held in focus by many humans. We can participate in this elevation of human conscious awareness by raising our own frequency with more wonderful thoughts and feelings, based in our awareness of our eternal conscious life force.

The force of unconditional love and joy is the most powerful force, because it is in alignment with the life-enhancing energies of the Creator. When we align our own imagination and feelings with this force, we express a destabilizing frequency pattern for the negative ones. Gaia is supporting us in this with her rising resonant frequencies, as is the entire cosmos. We are living in a very long wave pattern that is shifting electromagnetic polarities. We're moving from negative to positive. This affects our entire being. Fear disappears with intuitive awareness of our eternal Self. The vibration of love is all there is that is real. It is the vibration of Creator Consciousness. We just haven't known this. We let ourselves be tricked into giving our life force to our slave masters, and we have agreed not to opt out of this arrangement.

That agreement is meaningless. It benefits only the slave masters. Let's get real. We are sovereign and free Beings. We have ultimate control of our most powerful ability, which is to focus our attention and awareness. This determines our present vibration. Whatever is going on around us is immaterial. What matters is our own focus. We control the quality of our thoughts and emotions. This gives us infinite creative ability, within the limits of our beliefs about ourselves, which is how we got enslaved. We took ownership of many false beliefs. We didn't recognize the negative polarity of the energies involved. We just wanted the experience, and we lost our Self-Realization.

We have the ability to know our true Self. The connection is through our intuition. It is the awareness of our deepest Being, where our consciousness arises from. This is where we can know everything in each moment. It requires our clarity of mind and emotion, with no attachments or beliefs, just being present awareness. We transcend ego consciousness, which becomes irrelevant. We can become aware of more poignant and beautiful situations and realms. We can also just be present and aware without judgment in the face of conflicting energies. We can withstand all temptation to change our polarity or lower our vibrations, because we can just let them pass without our alignment with them. We can choose to align only with positive, high-frequency energetics. This contributes to the dissolution of parasitic, negative, life-diminishing influences.

Realizing Our Present Awareness

It can be a challenge for us to just be present in awareness and hold that focus, but it is a necessary step toward Self-Realization. In this state, we can be very sensitive to our intuitive knowing. We can feel our own vibrations, and we can choose to use our focus creatively. We can envision and feel ourselves living in the world of positive, high-vibratory experiences. We can be grateful for our conscious Being and for every situation that we experience, because it is for the benefit of our expanding awareness.

As we are able to stay high in our emotions and thoughts in threatening situations, we elevate the energy around us. We become inspiring to others, and our personal experiences come into alignment with our rising energy signature.

Without internal interference, we can align with our true nature. We have established our limiting distractions and beliefs over the course of our lifetime, and they are deeply-seated. We need to resolve them with patience, compassion and strong intention. We created our ego to be needy and to take us into low-vibratory, negative situations in order to expand our awareness of what those vibrations feel like. We've learned to survive as slaves, victims, villains and rulers, all without higher guidance.

When we've had enough negative experiences to know that we prefer positive, high-vibration thoughts and feelings, we can begin to trust ourselves to be true. With practice this awakens our latent abilities, which we could not realize until we were ready to live in the expanding awareness of universal consciousness, the consciousness of the Creator. This is our eternal present awareness beyond limitations, and it is the power of our creativity.

Getting Through Challenging Experiences

We are in this dimension for the experiences that we can gain to enlarge our awareness and deepen our understanding and compassion. After many lifetimes as humans, we have accumulated some deep fears, that can disable our ability to align with our higher nature. We have had to live in the limited consciousness of our ego without higher guidance, not knowing that we actually are eternal Beings of unlimited consciousness and unconditional love.

When we are in the grip of deep fear or hatred, even fearing for our life or wanting to destroy someone, we disable our ability to understand our situation and receive higher guidance. We can choose to accept our situation and truly experience it. When we are thoroughly imbued in the negative energy, if we can then take a few deep breaths, we can begin to regain our ability to be open to our intuitive knowing. If we can then choose to open our heart and feel compassion for ourselves for our deep hurt, we can begin the process of resolution. We can forgive ourselves for getting stuck in negative, low-vibration energy and begin to know that we can choose to open our awareness beyond our current state of feeling and thinking.

We can choose to become aware that we are much more than our current human consciousness and physical presence. We can ask for inner strength and guidance from our greater Being, our unseen guides and angels, and our divine consciousness. We can seek awareness of our inner knowing of expansiveness and eternal Being.

If we take advantage of our innate ability to choose whatever level of thought and emotion that we truly want, we can transform our lives. We can train ourselves to resolve all of the false beliefs, old fears and traumas, and our blocks to expanded consciousness. This is a process, and it may require a strong intention and much practice. When we resolve our self-imposed limitations, knowing our true, expanded Being becomes possible, because we no longer need to protect our ego consciousness. We can open ourselves to the awareness of the One we truly want to Be.

Transforming Our limitations

We are fractals of the limitless Creator Consciousness, and we are also personal holograms, in the sense that each of our atoms and subatomic particles contains our entire body in its design and in its mass, on a fractalized scale. We are also fractals and holograms of the entire cosmos, as is each of our atoms and subatomic particles. All energetic patterns in the quantum field are fractals and holograms that we can observe and make manifest by our recognition.

We are Beings of light. Each cell of our body constantly emits photons. These are conscious beings of light. Their presence is only pure electromagnetic waves that are self-luminous. Each photon is a hologram and a luminous fractal of our entire being. In our essence, we are luminous beings, but on a level below the perceptive ability of most humans.

In order to be able to have an authentic human experience, we have limited ourselves to a realm of low-vibration, negatively-polarized energy. All of our empirical, human experience is enclosed within the range of frequencies of conscious life on this planet. This is a small, limited, mostly negative, fear-based resonant energy. There is much more that we are capable of experiencing. To find out what that is, we can intentionally change our polarity to positive, with positive, high-vibration thoughts and feelings. This expands our awareness into a higher, more refined range of experiences.

As we mostly focus upon the kinds of experiences that we love and welcome into our lives, they become real for us. It works the same way in the opposite direction. What energy patterns we focus upon is our constant free-will choice. We have allowed our choices to be limited, but there's no cosmic requirement to continue to do so. We can orient ourselves to living with life-enhancing thoughts and feelings for ourselves and all conscious beings (which is everyone and everything). The essence of everything, from the most minute, to the exponentially greatest, is a conscious being. All conscious beings constantly receive the light of the Creator and emit photons. We regulate this flow with our thoughts and emotions. The more life-enhancing light we can open ourselves to receive, the greater our luminous radiance flowing into our lives' expressions.

We can become brighter, the more we align with our natural energetics. This is the pattern of life that is the essence of our Being and is constantly flowing to us as our conscious stream of life. We direct this stream of light with our focused awareness. Opening our awareness is an intentional act that can transform our traditional, limiting beliefs. We can realize intuitively how we can consciously extend our awareness beyond our physical world, just as we do in our dreams.

If we choose to be aware of the high-quality energy that we naturally can hold in our essence in every moment, we can intentionally transform ourselves by becoming transparent mentally and emotionally, knowing intuitively how to master every circumstance in life-enhancing ways.

Choosing Our Experiential Reality

What we perceive, imagine, feel and believe in each moment determines our experience. Each of these components depends upon what we choose or accept. We do not experience what we do not choose and accept as our personal reality. As humans, our empirical perceptive ability is limited to a small segment of the totality of the quantum field, but the potential of our conscious awareness is unlimited. We can open ourselves to our inner knowing in a way that is far beyond our human perceptive experience, and we can enhance our ability to experience life on this planet.

The primary factor in creating our life experiences is our belief structure. This is how we have established the limitations to our awareness. We cannot experience what we do not believe is possible. If we believe that we are shameful, guilt-ridden and mortal, we cannot experience our loving and immortal true Being. If we choose to transform our lives, we can resolve our false, limiting beliefs about ourselves, first be recognizing them and then by realizing their unreality.

How do we know what's real? The only reality is what we know within ourselves through our intuition. We are designed to be completely Self-determined. Each of us is the creator of all of our experiences through our inner processes, which are determined by our perspective. We have allowed ourselves to be taught by others about what to believe and what to expect. None of this has any meaning or effect upon us, unless we choose to make it so. Our free-will and creative ability extend to absolutely everything.

We are limited only in that we cannot interfere with the conscious Being of others, because we are all sovereign in our eternal Selves. We are fractals of the Creator of all, endowed with unlimited creative ability, as much as we trust ourselves to experience. This we can know only within our own Being through our intuition, which we can teach ourselves to know intimately in deepest love and alignment with Creator consciousness.

Participating in Creator Consciousness

The perfection of our Being is enhanced in every moment through our focus into Creator Consciousness. The most elevating vibrations of our Being flow to us constantly in the unlimited consciousness of the One, Self-Realized, all-conscious Creator. We participate in this consciousness as much as we allow ourselves, with no blocks outside of our own creations. Creator Consciousness encourages us to be completely open to our intuition. This is our connection with the divine, the pervading energy of positive, high-vibration life of unconditionally-loving connection with everyone and everything. Living in everything, consciousness creates the form and energy patterns of everything in every moment. Everything, from the incalculable minutest flicker of electromagnetic waves and patterns, to the vast cosmos and beyond, is created in universal consciousness, which envelopes everything in itself.

We are not alone. The infinitely-conscious Being is the ultimate Creator, out of Whom we arise as Self-Realized, personally-aware Beings. In order to have realistic experiences, we intended to participate in human life on Earth within the limitations of consciousness that we created and maintain. We can alert ourselves to these limitations, whenever they intervene in the divine flow of conscious life force through the heart of our Being. Here we can reach resolution, increasing our vitality and level of joy.

Through positive, life-enhancing feeling and envisioning, Creator consciousness sets us free to exercise our creative desires in alignment with Its vibrations, which come to us intuitively. We have free will in every instance, and we can choose our focus constantly. We can be aware of the vibratory level that we prefer, and we can direct our attention there. As is true of every creative moment, the specifics of form are less important than the vibratory level of our own state of being.

Once we have resolved our limitations, we can be optimistic, and we can open ourselves to positive, high-vibratory experiences. We no longer need to rely on our ego-consciousness for guidance, because we can know everything we want through our intuition, which is our inner knowing. Through our intention, we can learn to be intuitively aware constantly. It is our natural state of Being.

The Root of Our Being

If we search for where our conscious life exists in the present moment, we will find that it's everywhere that we have an awareness of being present. It's throughout our body and our energetic presence. It is in every cell and sub-atomic wave/particle. And it's in every dimension. Our consciousness is unlimited. It is the consciousness of all that exists and perhaps more. Our awareness is as great as our beliefs allow us to go, and it is potentially unlimited in alignment with the universal consciousness of the Creator. We can feel this alignment as joy and ecstasy.

One pathway to awakening to a higher realm of living, is feeling as much as possible the energetic level of joy, compassion and abundance, of being sovereign and eternal. Whatever personal limitations or fixations keep this state of being beyond our realization of it, can be resolved and released, freeing us to realize our clear presence of Being. We have an innate energetic consciousness in our personal Being, that reads our ego-conscious vibratory frequency and establishes a resonant frequency as our state of being.

Our energetic status consists of everything that each of our cells remembers about us from all previous lives. All of our unresolved negativity can be repolarized in moving from fear to love.

There are great challenges in this transition, requiring strong intent at resolution of everything we have believed about ourselves. It's all backwards and upside down, even our language.

Once we determine to be positive in the love frequency, our entire lives change, and we discover intuitive guidance in every moment. We can know and feel the energy of mastery in every situation. We can feel that we have no attachments, and we can always have an abundance in all aspects of life, if we do not limit it subconsciously or consciously.

Making our way through the emotional minefield we have inherited from our past is challenging. It might take lifetimes to resolve all the problems, but all we need to do is find our intuitive knowing, focus on it and learn to understand our inner prompting. We receive our guidance along with our conscious life stream throughout our Being and can be aware of everything through our alignment with the high-frequency vibrations of our intuition.

A Path to Inner Realization

When we are deeply in love in a relationship with a soulmate, we experience a state of being in joy that we want to stay in forever. This realization happens between two persons, and each of us is experiencing our own feelings, which our counter-part stimulates in us. We are actually in love with our own Being. Without a physical partner, we can still do this. We just need our imagination and emotions. We can be so in love with our essential Self, that we emanate this energy to everyone and everything around us in the quantum field. It is a high-vibrational energy pattern that can re-magnetize negative energies and enhance positive ones.

With this state of Being, we can transform others in alignment with Gaia, Spirit of the Earth. We can live without fear, because we know we are our eternal present awareness. We can express ourselves through any form we can imagine. Currently we're being human, but we are not required to limit our awareness to such a small part of our consciousness as the traditional spectrum of humanity.

We can intentionally expand our awareness. Whatever we want to do occurs for us, if we align with its vibrations and believe that it's real. That's how we create things. We imagine and feel what we're looking for, and the energy pattern then appears, first in our realization and then in our experience.

We can be open for anything at any time. We can look beyond our human awareness in our own consciousness. To be able to do this, we need to resolve our beliefs about ourselves that limit our awareness. The most important of these is the belief in our suffering and mortality. This is the root of fear and the spectrum of ego-consciousness. It is a negatively-polarized, low-vitality realm, but people have become comfortable with it for a limited life.

A way to realize being beyond a low-vibratory level is to push our imaginative limits into the deepest true love with a soulmate. The more we can experience the joy and ecstasy of this relationship, the better we realize a greater part of our consciousness. It can be an exercise in expanding ourselves, and we can learn to apply this perspective of love and joy in every moment with everyone and everything. It transforms our lives, as circumstances arrange themselves in alignment with our energetics, and it can be the beginning of higher consciousness.

Awareness Beyond Our Limits

All patterns of energy that confront us are filtered to our awareness through our beliefs in our limitations. We have the ability to filter and limit the conscious life force that enlivens us and gives us consciousness. Although they are deeply imprinted on our subconscious innate being, we are not required to hold these beliefs. They were all formed in the negative polarity of fear, with the intent of protecting our physicality without higher guidance.

As we enter the path of knowing our true Self, we can be observant and aware of our emotions in every encounter. They detect the quality of energy patterns that stimulate our feelings. Our emotions can also be creative when we want to feel them. As we can imagine scenarios in alignment with our creative emotional state, we can strive for clarity in aligning with the vibratory level of joy and equanimity.

The traditional human energy spectrum is negatively polarized. It is the realm of ego-consciousness, with its limiting beliefs in shame, suffering and death. These are real for us only because we believe in them. We experience the energetic frequencies of our energy signature, which is the expression of our predominant state of being.

When we decide to become positive in the vibrations of joy and compassion, it is because we know that they feel very attractive. We need to realize that this decision changes our lives dramatically in every way. It's a leap in consciousness. It may seem like a leap of faith in a power beyond the empirical world. It is actually just a personal choice in every moment to feel really good and expect that every situation we encounter will be a wonderful experience.

The intent of Creator Consciousness is to expand into new experiences of all kinds that enhance life everywhere. This is the positive polarity we can align with in every situation, including challenging ones. Our emotions convey the guidance of our intuition as well, when we learn, by our intent, to be aware of it. It is our intent that can transcend and resolve our beliefs, and that enables us to move beyond our limitations.

As we become clear of inner drama, we can begin to realize that our awareness is beyond the body. Our imagination and emotions are not in the body. Our awareness is also beyond space and time. Once we resolve and eliminate our limiting beliefs, we can be emotionally and mentally clear and able to know our intuitive guidance. We are part of universal consciousness, which enlivens and guides all persons and things. We can go deeper into universal consciousness, as our awareness expands greatly into feelings of goodness and prosperity. This is the quality of energy patterns that we attract in alignment with ours.

From Perceived Oppression into Transcendence

Few of us regard ourselves as truly free and fulfilled. That is part of the purpose of life here. We are to remember what true freedom is, and we can compare it knowingly to what is experienced on Earth. We are being asked by the enveloping energies of universal consciousness to make an important decision based on deep experience. We are being asked if we want to be free in a more life-enhancing way, in our memory of eternal awareness.

Our oppression is in our own belief. We have accepted the belief that we are mortal, and that we die. Anything that we believe threatens us with shortness of life instills fear in the consciousness we have accepted as our own identity. Take away our fear and repolarize it as love, and there is no threat. There is only a scenario in which we are genuinely joyful. This is how gaining a loving perspective works. It moves us into a positive, higher-vibrating dimension of life.

Negative experiences stimulate fear and are life-diminishing. Positive experiences stimulate many wonderful feelings. These two polarities co-exist for us, but we have been largely unaware of the real effects of the positive, because we have believed that we are limited beings, with many needs that can only be satisfied by others. We are not required to have these beliefs, since they are not true.

Our true Self is far greater than the human identity that we have assumed. We do not need to believe that we are eternal, because it is true. It is something that we inherently know. No one has to teach us. We just have to realize the unreality of our limiting beliefs and free ourselves to be our timeless Self-Awareness, able to express ourselves without limits. If we can let our imagination run and elevate our feelings to the level of joy and stay in this state of being for a while, we can have a memory of what we feel in our true Being.

We are our own oppressors. Fighting anyone else for freedom only aligns us with oppressive energetics. It is what turns the oppressed into the oppressor. This energy does not exist in the higher dimensions, because it has no life force from universal consciousness. It exists only in human beliefs, and it parasitizes our life force through our alignment with its vibratory patterns in a negative way. The way out of this situation is to change our perspective to awareness of our freedom and abundance. These energetic patterns exist all around and within us. Our intentional recognition and alignment brings them into our experience and enables us to live in a higher dimension here and now.

Realizing the Nature of Our Beliefs

If we choose to examine the essence of our beliefs, we find that they provide a compartment of consciousness that we contain ourselves within. While we believe that we are mortal, we are not open to realizing our eternal awareness. While we believe that we are our physical bodies and our separate identities, we cannot realize our presence beyond time and space, nor our multi-dimensionality. While we believe that we have separate minds and consciousness, we cannot realize our participation in universal consciousness.

We often speak of a veil that keeps us from awareness of other realms. That veil consists of our beliefs. We have intentionally limited ourselves in order to participate fully in this drama of humanity. Our ego consciousness distracts us as much as possible from questioning our beliefs. We have become comfortable with living in fear and do not want to be aware of the unknown. It is in the unknown, beyond our ego consciousness, that we can resolve our limitations and transform our fear into love.

Once we determine this, we can decide to open ourselves to expanding awareness beyond our empirical experience and into a timeless sense of Being. In this state, we can enter awareness of the consciousness of the Creator, the universal consciousness, filled with unconditional love for everything created constantly. Here is unlimited vitality and fullness of knowing and Being. Enhancing all of life, a feeling of joyous ecstasy pervades everywhere.

Our natural state of Being is in alignment with the feelings of the Creator, and our guide for this level of vibration is the spirit of our planet, Gaia. She resonates at increasingly higher frequencies and intensities. Her energy is expressed in all things natural on the Earth. Wild animals and plants are aligned with the energy of Gaia. Some creatures express this energy more than others. When we listen and feel the vibrations of their songs, small forest birds create a melody in resonance with Gaia, stimulating compatible feelings in us.

By having boundless imagination and feelings, we can sensitize ourselves to our inner knowing and feeling. Becoming sensitive to our intuitive guidance is something we can search for and practice. When we search for it, intuition is the immediate knowing of thoughts, feelings and motivations in every moment. Through the energy of our heart, it is given to us in ways that each of us can understand, if we want to. By our intention, we are guided into realization of everything we need to know and feel in order to express our conscious resonance, the quality of our state of being.

Our Possible Intentional Evolution

The great challenge for us on the inward path is becoming aware

Chapter 2. Resolving our Limitations

of all of the parts of us that we need to transform from negative to positive polarity. We are asked to transform every bit of fear and doubt into love and confidence. It's a matter of perspective. Whichever one we choose intentionally or by default, is the energy spectrum that we create for ourselves. As humans, we have been unaware of this process, and in our encounters, we have relied upon ego consciousness to react, often defensively or aggressively. In this state of being there is little awareness of intuitive heart-energy and guidance.

In our emotions, we possess a great tool for our expansion into positive polarity. Emotionally we know what stress and fear feel like, even in minute magnitude. We also can feel true love and joy. This is the difference between ego consciousness and intuitive awareness. When we become sensitive to our intuition, we can dissolve ego consciousness into intuitive awareness, because the ego is no longer needed as a survival guide. In the vibratory level of the unconditional love that we can realize that we are enveloped within, we can also intentionally realize our eternal presence of Self-awareness of unlimited Being and creative ability.

Each of us has the ability to awaken to unlimited awareness, in alignment with the consciousness of life-creation and enhancement. In our intuition we have a direct connection to universal consciousness. When we decided to play the game of human incarnation, we knew that we'd be locked into a compartment of our consciousness that we might not be able to get out of. We made our non-intrusive intuition the way back to our true awareness, but we have to choose to become aware of it in all its subtleties.

In order to truly know and feel our intuitive guidance, we can disengage from the drama of human life and just be present in awareness on a positive, high-vibratory level. Out intuition is beyond polarity and draws upon the experience of all lifetimes and ways of being. When we're focused on our intuition's loving and wise guidance, our emotions let us feel that we have arrived at a state of knowing. Intuition is the source of everything that we deeply know without any kind of proof. It is the first cause of knowledge and wisdom, and it ranges in all dimensions beyond the empirical. It is based on the unconditional love of universal consciousness, which we are part of in our presence of Being. In our essence, we are fractals of the Creator, and we are the same Being with all conscious beings. All have the light of conscious life force, which continuously emits conscious photons, tiny quanta of light, creating a radiant energetic aura. The magnitude of our aura depends upon the polarity and vibratory level of our chosen energetic alignment, and we can intentionally access our intuitive sensitivity to this radiance.

Awareness of Our Enveloping Energetics

Suppose, as much as we can, that we approach every encounter with an open heart, expecting to see the light in the eyes of all living beings. It is the conscious life force that sustains everything and enlivens all conscious beings with unconditional love. In this way we are creating experiences of greater love in our own essence. We can feel this, and it elevates us. In this state of being, we can resolve and transform any negative encounters, until they no longer exist for us.

We are all the same essential Being, existing within the consciousness of the Creator and connected in unconditional love. We've gone on an excursion into limited consciousness, but we can return to awareness of our unlimited Selves whenever we choose to drop our self-imposed limitations. Our limiting beliefs are negative and based in fear, ultimately fear of suffering and termination. When we change polarity to positive, everything is based in love and eternal vitality.

There is a conscious life force that envelops us with positive, high-vibratory energy, and this force is strengthening. We know this from our measurements of the vibratory patterns of the Earth. We are being drawn into greater Self-awareness. In unconditional love, we can intentionally bring ourselves into resonance with Gaia, Spirit of the Earth. This is the natural energy of the heart of our Being and is what we naturally express in our radiance, when we are in a state of joy and feeling fulfilled in every way.

As we can become fully open to our intuition, we can learn to know the inner guidance that we always have. We can be confident in our eternal, expanding Presence of Awareness, growing in joy and transcendent vibrancy. As we raise our vibrations through our imagination and feelings, many new and different experiences become available to us, and we attract others of similar vibratory levels. We are all being attracted to positive, high-vibratory living, and are being moved by the enveloping energetics toward greater alignment with life-enhancing thoughts and feelings in every moment.

Aligning with Our Innate Being

If we desire to make the leap in consciousness from negative to positive energetic polarity, we can gain the confidence needed by deepening our intention and compassionately working and communicating with our innate being, the part of our consciousness that holds our beliefs and our physical presence. Our Innate controls every cell and function in our bodies, and it possesses a deductive awareness. It is subject to the inductive awareness of our ego consciousness, which it tries to make sense of from a deductive perspective. The functions of our bodies are a result of the vibratory level of our Innate or subconscious self. It sees us as itself, but it doesn't completely understand us. We can help the process of alignment with our Innate by intentionally being compassionate, direct and consistent in the level of our vibrations. Our Innate is designed to internalize the energy of our state of being, resulting in the condition of our bodies.

When we are in alignment with our Innate, regeneration of our conscious selves results in regeneration of our bodies. Everything begins in consciousness and is held in a stable vibratory spectrum by our attention and energetic alignment. By changing the alignment of our focus, either physically or in our imagination, we change our experience. Our attention is the source of our creativity. Within our attention, we can calculate, visualize and have feelings. These all have levels of vibration that we are aware of. We can control our level of vibration with our intentionally positive perspective of joy and compassion in alignment with our Innate being.

By being loving within ourselves, we can resolve our internal differences and life-diminishing energies. Nothing, outside of ourselves, can enclose us in negative energy. Unless we want this, we can change our state of being by intending to open ourselves to higher inner guidance, and then paying attention and always expecting high-vibratory people and situations. As we continue to be in a positive state of being, we can learn to be comfortable and confident in being positive. We can be more compassionate and loving, because we can know that we can create whatever we need and can train our Innate to do this without our attention, except for our gratitude.

Gratitude is the one feeling that is necessary in a loving relationship with our own inner being. Once this relationship is in alignment, it's much easier to be mentally and emotionally open and clear. It becomes easier to stay positive and confident. Our awareness continuously expands as our vibrations rise, and we gain compassion and wisdom.

3.

Enhancing Spiritual Development

Aligning with Our Eternal Self

The better we feel, the more loving, the higher we go in our personal vibrations, and the better our lives become. Our state of being depends upon our love/fear ratio. As long as we believe in our mortality, we are always in fear of termination of our ego consciousness. The cure for this is an out-of-body experience. The next best thing is listening to those who've done it. That might be enough to allay our fear and instead, allow us to believe that we exist beyond time and space. Our beliefs are our free-will choice. They can be whatever we want them to be, but they are held deeply and need deeply significant encounters to hold them subconsciously.

It is our life force that enables us to create everything that we experience. We are absolutely sovereign in our personhood. Our personal energy signatures attract and interact with energy patterns that become our experiences. These are the energetics that can lift us out of fear completely.

Praying for something most of the time does set a strong intent and attracts experiences that align with our vibrations. When we're attracted to someone or something, we feel the energetic presence aligning with us. By feeling this presence, we can align with its vibrations emotionally. This frequency alignment provides strong life force to radiate our feelings into the quantum field, out of which the energy patterns emerge for our recognition.

It is our emotions that magnetically empower all of this and carry us into scenarios that are in the spectrum of vibrations of all of our thoughts and feelings. This is our energy signature, and it radiates its frequencies all around us. We can imagine situations in which we feel deep love and compassion. If we really go into this emotionally, our awareness begins to expand into higher vibrations and an understanding of unconditional, unlimited love. This is the quality of the conscious life force that enlivens us and connects us to universal consciousness as personal Beings.

It is from our eternal presence, beyond time and space, that we can be masters of our current situation. We have compartmentalized our consciousness in order to have this human experience, and now we have the opportunity to expand our consciousness as much as we want. We can begin to follow our intuitive knowing, which flows to us through the energy of the heart of our Being, our expanded Self. As we practice this awareness, we increase our sensitivity. Developing this sensitivity and alignment with our inner knowing allows us to live in the high vibrations of joy and abundance.

Our Greatest Challenge

On the journey into Self-awareness, we may confront many

issues of low vibration that require our attention, which we may give them in compassionate understanding. As long as we have any attraction for their attention, they will be in our experience. By choosing to live in the realm of love, we can leave all fear-based beliefs and attractions of low-vibration energies and experiences. The greatest attachment to fear is the belief in our mortality. Watching Thanatos TV on YouTube should help to transform that belief. Many people have died, came back and have reported about their experience. When they died, their conscious presence of awareness just expanded. They became aware of themselves beyond the physical body, and some were able to enter universal consciousness.

Experience is the best teacher, but for those of us who have not died and come back in this lifetime, or have not had other out-of-body experiences, the issue of our mortality is a difficult emotional knot to resolve. We have to move beyond hope, but how can we become aware of our own eternal Being? This is the ultimate issue for us. We can begin by intending to know our true Self. We can imagine what unconditional love feels like in relation to the persons and pets that we're close to and whom we truly love. This is the vibratory level we can stay in, while continuing to expand our awareness.

Our emotions are not bound to time and space. We can feel energies that are invisible and imperceptible to us. These feelings are in the subconscious of all of us. To become aware of them consciously, we can open ourselves to our intuition. This is where we know everything that we know, apart from any influence outside of our own inner awareness. Once we intentionally connect with our innermost intuition, we can become aware of our guidance for living in every moment. This guidance transcends personal beliefs and perspectives not in alignment with the life-supporting energies of universal consciousness. When we are perfectly attuned to our intuition, we can always know the truth about everything, and we can feel the vibration.

Participating in a Higher Quality of Life

The Spirit of our Earth, Gaia, has been vibrating at very low vibrations for a long time, and it nearly killed her physical body. She has done this for the sake of humanity, so that we could explore the feelings of low vibration lives, based in fear of every kind and intensity. We and Gaia are now the receptors of large transmissions of gamma-ray photons, which are creating more light spiritually and physically. We are now constantly enveloped in these waves of high-frequency energy, and are being urged to expand our consciousness and become aware of our true potential, because the lower vibrations are not aligning with Gaia and will soon lose their amplitude and disappear into unrecognizable energy patterns. They are already becoming unstable and insane. They are being recognized as never before.

Gaia is regenerating. As we continue to align ourselves with her rising resonant frequencies, we find ourselves living more fulfilling lives, because our rising frequencies attract energies of the same quality of vibration. Our perspective can become more loving and compassionate, because of our experiences in the low vibrations. We could never have known this kind of energy in any other way. We know how enchanting, addictive and destructive the low vibrations can be. We can forgive every conscious being for giving us the experience of living in the frequencies that act destructively in reducing our life force.

The spectrum of energy that humanity has lived within for eons is an electromagnetic creation held in mental and emotional focus by all of humanity. We agreed to limit ourselves to this spectrum of energy in order to have this human experience and learn to rise beyond our limitations from within them. To make this transition, we can go into the feelings stimulated by the heart of our Being. These feelings come from our true Self. They are high-vibration feelings aligned with love and compassionate understanding. If we intentionally open ourselves to become aware of them, they will come into our awareness, as well as the qualities of the situations that we are constantly creating.

Along with the emotional feelings coming from the energy of our heart is our intuition. This is how we innately know what we know. Intuitive guidance is always present. Part of it is our conscience. We have to open ourselves to become aware of it. It may come as words, symbols, pictures, energy running up and down our spine, or any way that we personally can recognize. It could be indescribable inner knowing. Once we recognize this energy, we can know what it feels like. We can align ourselves with its frequency pattern to create situations for ourselves. The resonant patterns of energy that we intuit can feed our imagination and feelings, radiating into and manifesting out of the quantum field as personal experiences of the same quality as our state of being.

Identifying with the Song of Our Heart

In a world going crazy with chaos and intolerable mandates, violence and destruction, we are being guided to redirect our attention to what truly matters and can lift us above the low-vibratory energies engulfing our planet. Only by holding our own energy signature in balance in higher frequencies can we change the energy level of humanity in alignment with the rising resonance of the Earth. We have the ability to transform the vibrations around us into feelings of kindness and love.

When we feel the vibrations of our heart in any situation, we are moved to acceptance, forgiveness, understanding and compassion for all who are involved in the drama of human life. Because we are the Creators, we can make ourselves exempt from the influences of the low vibratory conditions by transforming everything we perceive and feel that has a negative polarity of fear and life-diminishing energy. We do this by intentionally changing the energy within ourselves to positive, life-enhancing vibrations in our own emotional and mental state. We can use our visionary and emotional abilities to focus on the light that is present in everyone we encounter.

Every human has a heart, and we can stimulate and enhance its radiance by our own transforming love and compassion. Every heart sings of joy and unconditional love unceasingly and forever. It is the conscious life force flowing from the Creator of all. We can call this energy forth in everyone we encounter through our own love and compassion, as we remain steadfast in our perspective.

Everything that exists is a manifestation of flowing patterns of electromagnetic energy in the unified quantum field, and we are connected to all of it through our emotions and mental processes. We constantly modulate the energy patterns in our presence by how we feel and think. By choosing to live in high-vibrations of feeling and perceiving, we can be motivated constantly by the vibratory patterns of life force flowing through our heart. In this state of Being, we do not attract low vibratory experiences, because we are not in alignment with them. We attract only what aligns with our vibratory patterns.

Just Being Here Now

We can begin by being in natural places, apart from humanity, places of beauty and majesty. Here we can feel the vibrancy of Gaia and her creative imagination. We are her guests here, and our intention can be to enhance our environment with wonderful feelings and deep connection with all of the conscious entities around us. We can be at peace, in serenity and just be aware. Breathe deeply, nothing more. We can be grateful to Gaia for providing this planet for us and for enlivening us, regardless of how we treat her body. She is waiting for us to align with her rising vibrations, so that we can all enter into interactions of kindness, joy and support.

As we align with her energetic patterns in the spectrum of joy, compassion, and love, we can feel that these are life-enhancing vibrations. We can learn to stay focused within this dimension of high vibrations. We can become aware beyond thought and emotion. Just present awareness. This is our eternal Being. It is Who We Are. This state of being comes with clarity. This is where we can become aware that we are within universal consciousness. This is our Creator essence. It is the connection through our conscious life force with the Source of our Being.

This level of consciousness of the Spirit of the Earth is attracting us to align with her, because it feels wonderful. Soon the resonant vibratory level of Gaia will be in a higher-dimensional frequency, where there is only love. Anyone not aligned with this will be too uncomfortable to be on this planet.

In the high-frequency vibrations of our thoughts and emotions, we enjoy greatly expanded awareness that we are not capable of in the current human consciousness spectrum. To expand we must leave behind all feelings of inferiority, shame, regret and every kind of fear. Only love and its vibratory mates can be attractive to us. There is nothing special we need to do in this life, except keep our vibrations high, which creates a fulfilling life. By interacting in love and compassionate wisdom with everyone, we become radiant masters of this energetic realm.

Participating in the Magic of Gratitude

The first step into higher consciousness is gratitude. When we are grateful for our consciousness, our vitality and our very Being, we are expressing high vibrations in alignment with the love of our Creator. The more we are thankful for, the more we experience situations that stimulate gratitude in us. The vibratory level of gratitude creates energy patterns that manifest as wonderful experiences, opening our awareness to greater expressions of our Being, our Personhood.

A grateful heart is a loving heart, filled with joy and well-being. This perspective can magically lift us out of tragic and negatively-polarized situations, transforming our encounters into greater awakenings into deeper love and compassion. This is the alchemy that we are capable of. We can transform poverty and desperation, as well as subjection to tyranny, into experiences of abundance and freedom through our focused intention of gratitude in alignment with the rising resonant frequencies of the Spirit of the Earth and our enveloping cosmos.

Our entire environment is flowing into alignment with divine intention, and we have the opportunity to intensify this trend with our own consciousness. As we practice gratitude in all areas of our lives, we elevate ourselves as well as everyone around us. This is our gift to our Creator and to all of life. By being constantly thankful for everything in our awareness, we can intentionally open ourselves to greater expressions of love and beauty. In this way we are the creators of a better world of gratitude and fulfillment for all.

Developing Our Creative Potential

Infinite consciousness is available to us in every moment. This we have not believed, and our connection has become dim. We have partially cut off the flow of our life force, which has caused our vitality to drop. We have aligned our vibratory level with negative, low-vibrational consciousness. We have lost our Self-Realization to the human spectrum of experience, because we have not believed that we can open our awareness beyond the energetics of this realm. To move beyond, we can take an inner journey into our deepest consciousness.

Deep realization and understanding happens when we open our awareness and drop our attachment to our physical body and everything else. We can recognize ourselves as pure self-conscious awareness. In this state, we can express ourselves as pure potential, life-enhancing vibrations. Our intuition is our point of awareness of our infinite consciousness.

If we can be aware of the polarity and vibrations of everyone and everything we encounter, we can become more sensitive to our own intuitive guidance. Our intuition will align us with the polarity and frequency patterns of our true Self. With attachments and beliefs, this cannot happen, because they overwhelm our intuitive knowing and greater awareness. Once we resolve our self-imposed limitations, we are free to live loving and compassionate lives in abundance and eternal, infinite knowing. We become masters in every situation, because we can intentionally align ourselves with the conscious life flowing out of the Creator. This is the Source consciousness that creates our body and enlivens us. It is creatively and unconditionally loving, while expressing all possible vibrations and vibratory patterns of all of the electromagnetic waves enveloping us. It is awaiting our recognition and modulation through thoughts and emotions.

In every moment we are the creators of our lives. We have been mostly unintentional creators, primarily from our subconscious, but we can change that, once we recognize our true identity and abilities. We enter the higher frequencies, because we have had enough experience in the lower ones to know that we prefer love over fear. We can intentionally align with kindness and compassion. When we seek our intuitive guidance, we become aware of it, because it is always present. This is the path for our own inner journey to infinity.

Enhancing the Vibrations of Our Personal Potential

As we stretch our imagination far beyond our current life experiences, we can enter the realm of greater light and joy. We can imagine being in alignment with the conscious life force streaming into us in every moment. This living force field enlivens every cell of our body and surrounds us with a living radiance of photons. This plasma stream is filtered through our beliefs about ourselves as it flows through us and gives us our expressions. It illumines for us the energy of our personhood.

Each of us has a unique perspective that radiates itself throughout our consciousness and expresses itself as our energetic signature. Through our perspective, we either free ourselves or enslave ourselves. In each moment we have the choice of which polarity and frequency patterns we want to focus upon. If we choose the negative, based in fear, we enslave our compartmentalized consciousness. If we choose the positive, based in love and joy, we free our awareness to enter the Consciousness of the Creator, the universal consciousness that constantly creates, dissolves and recreates everything with the intention of enhancing life everywhere.

We can become the intentional creators of our lives by mastering our thoughts and emotions. We can begin by being aware of their changing quality in every moment. We can take our awareness into a dispassionate observer perspective, resisting nothing, supporting nothing, but being aware of our personal ego-consciousness operating with free will throughout the human experience. It is being impartially aware of our ego-awareness and operations. Our ego will display for us all of our limiting beliefs. We can penetrate the ego's defense of its hidden emotional knots and put-downs, through our intention to know the truth about ourselves.

With compassion and wisdom, we can clear the negative patterns that have enslaved us within the world of humanity. Shifting to a positive polarity in emotions and thoughts is a major life change and can sometimes feel as if we just got reincarnated and need to learn how to live on this planet. It means that we can realize unconditional love, in expanding consciousness that includes everyone and everything, we can know and feel that we are completely cared-for on every level of Being, and we trust our intuitive knowing implicitly. We can be aware of our intuitive prompting in every moment. It is always instant, as soon as we focus on something or someone. It may be dim at first, but as we give it our attention, we can become acutely aware of it. It is our intuitive knowing that can guide us truly on the path to Self-Realization as a Being of light and love and infinite creative ability.

The Divine Purpose of Evil

The nature of evil is destruction of our life and vitality, usually through pain and suffering. Evil separates us in our consciousness from the Source of our life and keeps us from experiencing the true essence of our Being. Evil is parasitical toward all life forms. It is part of an imaginary world that we have created and maintained through our conscious awareness, which feeds our life force into negatively-polarized-and-charged, low-frequency energetic patterns of experience. The deeper into evil that our focus can take us, the more sinister and torturous it becomes.

We have had lifetimes dedicated to thoroughly experiencing the depths of evil. In our real Selves, we could never experience or even want to imagine such energetic patterns. As Self-Realized beings of eternal, free-will awareness, we could not take threats seriously, because they're too obviously illusory. We wanted to deepen our experience in all ways, to expand the presence of love and compassion. That meant having to experience seriously living within a negative, low-frequency band of vibrations. We now know that experience thoroughly. At some point, we're ready to move on to better things.

Meanwhile, we can be focused on realigning ourselves with Creator consciousness. Dealing with evil is a necessary element here. In the illusory world of humanity, we can be successful in this by intentionally aligning our focus of attention with our intuitive knowing. By delving ever deeper into our intuition, we can come to a full realization of who we are in a higher-dimensional reality. It is here that we can recognize the unreality of the negative realm. Its energy patterns have no divine life force, except for our own contributions through our shame, guilt and fear. We felt all of this in order to deepen our compassion and understanding in the expansion of our consciousness. We now know what we, in our expansive Being, could not have been capable of knowing. In our unlimited creativity, we have a deeper understanding of how to use our attention and focus.

Working Successfully with Our Beliefs

Our limiting beliefs exist for the protection of our ego-consciousness, so that we can have a convincing human experience. The ego knows it is mortal, and it doesn't trust higher guidance to be true or understandable. It lives with attachments and fears. It doesn't believe that we can live in love and joy without limit.

For eons humans on this planet have believed that we are separate physical persons. This is how we learned to recognize ourselves. It is a belief structure that we can resolve and transcend, once we realize the energetics involved. We can live in alignment with love and joy, which exist in a spectrum that is positively polarized and of high frequency vibrations. We can intentionally be in a state of alignment with the most wonderful feelings and thoughts. All of this is beyond ego-consciousness, by which we have believed that life in a higher dimension is not possible, while we are in this incarnation.

A higher dimension of living is present for us right where we are. It is a state of being. No dimensions are places in space and time. They are complex energetic patterns that we can recognize and make real for ourselves in our experience. The empirical world is a compartment in our consciousness, held in awareness through our beliefs about it. If we intend to recognize a higher dimension, we can do so with our attention, as we imagine and feel what it would be like. That is the energetic vibratory pattern that we can intend to recognize and accept as real. We can learn to direct our attention to energetic patterns that feel good and fulfilling, where compromises don't have to be made.

In order to realize our limitations, we can intend to recognize a higher dimension. We can be in a serene, comfortable environment, in which we can leave behind the concerns of life, and we can take some deep, rhythmic breaths. Here we can relax and just be present in awareness. Our awareness is not limited by space and time. It is quantum and eternal. We can imagine and feel wonderful in alignment with the energies of Gaia. As we open ourselves to positive, high vibrations, we can learn to be sensitive to our intuition, our higher guidance that is always present. This can become our state of being, enabling us to live and be aware inter-dimensionally.

Greater Realization of Our Self

We are here to create a reality that is a higher dimension of human experience, a positive polarity, high-vibration environment on a regenerated Earth. Only with our conscious intent can we do this. We can learn to be our true Selves, the higher version of who we feel we are. By opening our awareness to joy and love, as much as we can, we begin the process of personal transformation to a fulfilling life. We are created to live in joy and ecstasy and to be guided in all our ways by our intuitive knowing.

Coming through the energy of our heart, our intuition is a constant guide, even when we're unaware of it. Once we recognize it, however, we can realize that its guidance comes from a Being of infinite wisdom and love. It is what we truly know deep within. When our perspective is beyond fear and personal drama, our intuition is the first feeling and thought that we receive in every encounter.

The apparent flow of conscious life force is toward the enhancement of life, in contrast to the death-spiral that humanity has allowed to transpire on our planet, due to our negatively-polarized consciousness. We can change this in our personal lives with our intent to rise into a higher-vibratory, positive life of compassion and joy. The vibratory level of our conscious state of being becomes our personal energy signature, which radiates its energy into the quantum field for manifestation in our experience.

Our conscious state of being attracts energetic patterns that are in alignment. Because our innate being has been compromised by our ancestry and training, we need to choose intentionally to be positive in every moment, and to transform our negative perspective of limitation and fear. With our intentional choice, we can train our innate subconscious to resolve our limiting beliefs and open our awareness to joy and intuitive knowing.

Once we are in a state of contentment, we can realize that our Presence of Being is our awareness, which exists everywhere and is beyond time and space. Consciousness is One, and we have the ability to know its entirety, beyond the compartmentalization of humanity's world. Our awareness is potentially unlimited.

Expanding Awareness through Intuitive Guidance

Our intuitive guidance can come in many ways, but if we are sensitive to it, we always know what is best in every moment. It can create an encounter for us, something that can give us greater awareness. It can inspire a feeling and visionary experience that is beautiful and refined. It can be words or mathematical or astrological symbols. It will always be in terms that we can personally understand, if we're open to expansion. It can inspire us with miracles and lead us to control of the quality of our physical world. Our intuition is part of our consciousness. It is guidance from our eternal, expanded Self and the consciousness of the Creator.

In our individual essence beyond the empirical world, we are infinitely and creatively aware. We energetically interpenetrate the awareness of all conscious beings. We can be aware of the awareness of anyone, because we are all the same Being in our essence. We are all constantly created out of the consciousness of the Creator. We are the Creator personified as each of us.

We are in a stream of conscious life force that has a resonant frequency that we can align with. This is the long-term wave pattern of human evolution and its guiding energetics. Our natural direction is toward greater light and joy. We can follow a different path, if we prefer, but with the flow of life that is enveloping us, we are being prompted to align with life-enhancing mental and emotional energetics.

We have the potential of creating our lives in a higher spectrum of energy. We don't have to change anything physically in our own experience, but we can transform the energetics, because all dimensions are present for us, as we become able to recognize them. Through our perspective, we recognize them and express our energy signature, which vibrates with the polarity and frequency of our thoughts and feelings.

By intending to recognize a higher quality of life in every circumstance and encounter, we can become aware of our inner light and the light in every being we encounter, even the deeply negative ones. We can intend to be in gratitude, compassion and love in every moment. This can be our perspective, allowing us to drop our limiting beliefs, for they are irrelevant in the eternally-present moment. Our intuition gives us our higher guidance, if we pay attention to what we truly know in every moment.

Expanding the Energy of Our Heart

Everywhere on our planet the heart symbolizes love, usually romantic love or love as an emotional attachment to someone. In the spiritual community the heart symbolizes divine love, an understanding of unity in spirit with everyone and everything. The heart is our connection with expanded consciousness and greatness of Being. Its energy pervades our entire body and awareness with the enlivening consciousness of the Creator. It is constantly devoted to enhancing our life, regardless of how we treat it. All of this is true for us on every level of our being—physical, emotional, mental, etheric and spiritual. Our physical heart is the manifestation of the heart of our Being, our connection with the Source of our conscious life, which we share with all beings.

We have learned to constrict ourselves, to limit the energy of our heart, which has resulted in lowering our vitality in order to comply with our enslavement to negative forces. This imposition is now ending, along with the rising energetic vibrations of the Earth. The polarity of humanity is turning positive. We are realizing our freedom to express our innate abilities in expanding creativity and unity in spirit. We can learn to gaze into each other's eyes and see the light of love within, and we can interact in joy with this spark of love, which is always present in every conscious being, even the dark ones.

Our heart constantly shares its life force with all parts of our being. It knows that its life force is unlimited and flows through it constantly from our Source, unless our self-limiting ego-consciousness destroys it. And so it is with our higher Self-consciousness. When we discover that our creative ability is unlimited, we can choose to give our creative love without limit. This is how our true Being expresses itself.

In our true Self-Realization we feel fulfilled in our expressions of love for everyone and everything, because we know that our sustenance comes from within our own Being. The more we open the flow of love through us, the greater our awareness and ability expands, and the greater our self-fulfillment manifests in every way.

Traversing the Inner Path

Our real ultimate limits are our imagination and our feelings. How far out are we interested and willing to go? How much beauty, enlightenment and ecstasy can we go for? There are no limits beyond ourselves, unless we design something like our current human experience and limit ourselves in order to play in this dimension. If we can imagine being free and in love with all that surrounds us and fulfills us within, we can create that state of being for ourselves. This is exercising our mastery of the dimension we focus upon.

Unless we have some energetic knots that need to be resolved, we do not experience challenging situations. If we can face our limiting beliefs with conscious, objective penetration, we can know what is true. Our Innate Being knows this. With gratitude we can open ourselves to our limiting energy patterns by recognizing their vibrations. We can naturally feel and imagine what they are, when we search for them within our awareness. When we realize that our limitations serve only to keep us from Self-Realization, we can resolve them, and we gain our freedom.

As we grow in loving, our level of joy rises and we are in ecstasy as a natural state of Being. This is how we raise our vibrations into a higher dimension of living. We can just stay as much as possible in ecstasy, and our lives magically elevate into positive, loving experiences. We can begin to encounter ecstatic versions of others, and there is so much love and joy. This dimension is right here, right now, by adjusting our perspective to align with infinite awareness.

To make the leap in consciousness that elevates and expands our awareness, we can strongly intend to be in a state of joy, and then keep expanding it as a normal state of being, raising the positive polarity and vibratory frequency of our energy signature, which is the expression of our subconscious-innate being. When we want to expand our consciousness and elevate the vibratory patterns of our lives, we can be aware of our innate knowing, our intuition. Its guidance is always present at the level of joy and confidence, and it instills gratitude in us. We know when we have found the heart-warming knowing of our true Self.

A Strategy for Awakening

As we progress on our path to expanded awareness, and after we have allowed ourselves to go beyond the empirical world, we find that we can live as the same characters that we have believed we are, and that we can change our character at will. If we can resolve our beliefs about ourselves and just be aware of our inner state of being, our level of vibration, we can penetrate into our inner knowing.

Everything that happens is designed to challenge us in ways that can help us to open more to awareness of our true Being. Often we need a shock to stir us out of the human hypnotic trance. Once we are free of our self-imposed limitations, we can be unlimited in awareness. Until we experience universal consciousness, we can practice accepting every encounter in joy and gratitude, seeing everything while in a state of infinite love as great as we can imagine.

In the consciousness of unlimited awareness, we can continue to play the human drama, but it's different. The lessons are on a higher level. If we want, we can even rise above all obstacles, just by being focused within a positive, live-enhancing perspective. Fear disappears for lack of our life force. There is only the vibratory level of joy, gratitude and love, recognizing and feeling the essence that we all share within universal consciousness.

We can act out of our awareness of unlimited supply of everything. It all manifests out of the quantum field of all potentialities, in the vibratory spectrum of our conscious and innate alignment. Our thoughts and emotions are carried on electromagnetic wave patterns that radiate into the enveloping quantum field. Electrically and magnetically, we attract experiences that align with our polarity and frequency patterns. How we are and how we react determine what we experience, regardless of the outward circumstances. To elevate our lives, we must elevate our mental and emotional perspective, until we are aware of being within universal consciousness.

Transformation through Laughter

Some of us have locked ourselves into a negative emotional state and are experiencing continuously diminishing energy, leading to depression and hopelessness. In this situation, we are closing ourselves off from realizing our inner light and the fullness of our life force. We get hypnotized by our acceptance of fear. If we can recognize that we are in this state, and that we have the choice to expand our emotional focus, we can enter into a world of positive polarity, where everything enhances life. There's no requirement to stay in negative, life-diminishing emotional limitations. We have to feel our way into positive polarity while intending to be aware of our eternal Self.

At some point we become aware that the entire human drama is a compartmentalization of our consciousness. We're here for the complete experience of this level of consciousness. Now we know what depression and hopelessness feel like, and it's time to change character by becoming the way we want to be, leading to a change in polarity to positive, regardless of what our outer experience may be. We can move from the realm of fear to that of love and joy. This is a challenge for our imagination and emotional state of being.

One way of freeing ourselves from our emotional trance is to laugh. Just laugh out loud. We don't need a reason even to feel stimulated to laugh. It doesn't matter how we start. We can just laugh deeply and loudly and keep laughing. This stimulates us to want to laugh more, and our level of joy rises as we keep laughing. This creates infectious energy all around us and gives us a positive polarity and perspective. It's essentially emotional shock treatment, and it feels really good.

Once we are emotionally unstuck, we can ultimately become unlimited in our awareness and creative abilities. We have been held back from our full awareness by fear, which has no essence of its own. It exists through our alignment with its energetics, parasitizing our life force for its manifestation. By emotionally changing to positive in compassion and love, we regain the life force that we had given to fear and depression. This can be a life-transforming leap in consciousness into present-moment awareness continuously. We can free ourselves from alignment with depression and all negative energies and transform our lives into the best that we can imagine.

Experiencing Our Multi-Dimensionality

Our empirical hypnotic trance is powerful, because we have accepted beliefs that make it so. Even though the dimension that we live in as humans is a small compartment of our consciousness, it fills our entire awareness, while we are embodied within it. We have adopted beliefs about ourselves that severely limit awareness of our own being and our energetic environment. We are aware of a limited spectrum of electromagnetic waves and patterns within the quantum field, which contains every possible energetic configuration and experience in all dimensions.

We have no innate requirement to live within our inherited and programmed self-limitations. We can open ourselves to our greater reality by intending to do so and searching for the qualities of the most life-enhancing thoughts and feelings that we can imagine. These are the qualities of true love, joy, abundance, freedom and eternal Being. They manifest naturally in our experiences, when we align with the feelings of beauty, generosity, compassion and deepest love.

Chapter 3. Enhancing Spiritual Development

To enhance our deepest understanding of life, we have come into embodiment with humanity in the spectrum of deeply negative experiences, but we do not have to continue to participate in these energetics. We know what they feel like, and we can transform our experiences with our imagination and emotions. We have to establish a high-vibratory state of being.

Because we are multidimensional, we can be aware of a deeper reality beyond our physical Earth-human experience. By adopting a perspective of compassionate wisdom, we can resolve our self-imposed and accepted limiting beliefs, freeing ourselves to realize the truth of our Being. We can know the consciousness of our Creator, because we are the same eternal and infinite Being. We are fractals of the Whole of universal consciousness.

As we practice opening ourselves to our greater Being, we free ourselves to experience expanded consciousness. Regardless of the world around us, we can intentionally choose serenity and joyful laughter as our state of being, and our outer lives change themselves for us to reflect the frequencies of our own consciousness.

The Home Stretch

We have incarnated in the most significant time on our planet, the final contest between the polarities of light and dark, the positive and negative. For eons humanity has been enslaved to the forces of life-diminishing energies in order for us to know the true experience of these energetic qualities. We have developed deep compassion and understanding as a result. The cosmic energies have shifted, and the Spirit of the Earth is rising in vibration, enhancing all of life here. We are awakening from the hypnotic trance that humanity has inhabited for eons and realizing the nature of the dark illusion that we have given our life force to.

The world is growing brighter, even as we face seemingly

increasingly destructive forces. These forces no longer have much energetic magnitude, and we can increasingly easily transcend them through our focus of positive attention. By searching for the energy of the heart of our Being within, we can find our divine Source and intuitive knowing. This is our pathway to alignment with the unconditional love and joy of our true Being.

We can intentionally feel in alignment with the most fulfilling vibrations, and we can do this as much as we choose. Imagining and feeling joy and wonderful scenarios by pure intention expands our personal positive state of being and brings us into alignment with our intuitive knowing. Our intuition has no boundaries. It is completely life-enhancing in every way. To align with our intuition, we can intend to be pure of heart, compassionate and loving in every moment. This level of vibration is not touched by negative energies, because they are on opposite electrical and magnetic polarities, repelling each other's alignment. As a result, our lives become naturally joyful and free, as long as we stay positive.

Being positive means living without limitations. It means refocusing from life-diminishing energies based in fear, to life-enhancing energies based in love. The mental and emotional vibrations that align with life-enhancement are the quality of our intuition and the energy of our heart. By recognizing and aligning with this level of resonance, we can transform our lives, regardless of what may be happening all around us. By following the energy of our heart, we can enter a higher dimension of Being.

Realizing Our Intentions

If we desire to expand our awareness into our greater Being, we can resolve our limiting beliefs with our focus on our eternal personal presence of awareness. Through our presence of awareness, we can resolve our belief that our awareness is limited to our senses and physical presence. We can intentionally open ourselves to knowing our eternal Self. If this is our desire, and we continue to feel higher vibrations, our awareness intuitively opens us to the experience of a higher reality without limitations. We experience a higher power of creativity and feelings of joy and gratitude.

Our return to Self-Realization comes with constant higher guidance through our intuition. We can be extremely sensitive to our feelings, so that we can know the quality of energy in our presence. It's most beneficial for us to align with the vibratory level of our intuition. Much is communicated here to our entire personal being. The more we desire to know in greater fullness, and we continue to be open to greater awareness, we can live in a natural state of compassionate understanding and freedom.

With our discovery of each limitation, we can decide its fate in our consciousness. If we resist limitations, we strengthen them by aligning with the quality of their energy and giving it our life force through our attention. If we accept our limitations, we can transform them with our focus on the high vibrations of eternal love and feelings of life-enhancing energies.

We are our present awareness, unlimited by time and space. We are eternally Self-sufficient, and are in intentional and unintentional control of our lives by the level of vibratory resonance of our dominant state of being. This is mostly how we feel in every moment. The more elevated we are, the more wonderful our experiences become, until we are living with miracles and wonders. To achieve this, we can use our imagination to project love and gratitude throughout our being, staying focused on goodness and compassion.

Potential Utilization of our Free Will

Once we learn that we are the creators of the qualities of our lives, we also realize that we can improve our lives by intentionally being joyful and playful. Whatever mental and emotional level we are vibrating at creates a resonance for our ongoing experiences. Living joyfully creates joyous experiences.

We are created to be able to choose to focus our attention and express our feelings freely in life-enhancing ways. We can learn to control our focus of attention in every moment. We can intend to be open to all positive, high-frequency energies and to be clear of limiting beliefs about ourselves. We can learn to just be present in awareness. We can allow thoughts and emotions to pass through our awareness, while we maintain a state of compassionate wisdom and perhaps having fun. Our focus and alignment with the energy patterns is what we get back as experiences.

Our focus can include intentional awareness of the eternity of our Being, our presence of awareness. From this perspective we can choose to be compassionate and kind in all encounters, because we control the energetic levels of our thoughts and feelings, regardless of outside stimulation. Responding to threats and intimidation, from a perspective that cannot be intimidated or threatened, we can successfully remain in a higher octave of energy. While we are in a positive, high-vibratory state of being, negative energy disappears into our former dimension.

We can experience very negative, threatening encounters, while maintaining a perspective of gratitude and compassion for all that is happening. Interfacing with the light at the heart of all we encounter, we can align ourselves with the consciousness of the Creator and enter a realm of amazing beauty and joy. It is all available to us to share with one another.

We are the creators of our new world through the rising frequency resonance of our energy signatures. While we pay attention to the present moment, opening ourselves to feelings of the energetic patterns around and within us, we can always know our eternal presence. Our progress in expanding our consciousness is a process of intentionally choosing to be positive, open, clear and present in every moment.

Chapter 3. Enhancing Spiritual Development

In Seeking a Higher Quality of Life

We have been shielded from knowing our whole Being by our personal beliefs and by distractions. Human society has closed its awareness to the energies that are natural to us and to the Spirit of the Earth. Many seekers of enlightenment have chosen to live in a natural environment, mostly apart from society, in order to align with the energy of Gaia. Realized alignment with our Earth Spirit can be calming and attractive for a positive life surrounded by beauty.

Nearly all of our life experiences have been distractions that have kept us in the human hypnotic trance of limited consciousness. When we desire to awaken from this trance into a greater reality, and our desire becomes intention, we are guided from within and all around in symbolic ways that we can understand. The vibrations of this guidance radiate through our intuition, from within the energy of our heart. When we become sensitive to subtle feelings and promptings, we become able to follow our intuitive knowing. It is always positive and life-enhancing.

Through our intuition, we can understand that the human drama requires our attention for it to be real for us. Everyone we encounter is a reflection of our own vibratory level, including the aberrations. Through our attention and energetic alignment, we create the human experience in physicality. By withdrawing our attention from human life as we know it, we can intentionally align with Gaia and live in gratitude and joy. We can greet every encounter with a realization of our inner light in the energy that comes through the heart of our Being.

Depending on our sensitivity to our intuition, we can be guided as much as we need, even in our predicaments. Intuitive knowing is always present deep in our consciousness, and it always depends on our reception for our realization. To receive it, we must practice being mentally and emotionally open, clear and present. By following our intuition, we can follow our prompting to be more open and unlimited. We begin this awakening when we want to be positive and bright in every moment.

Inner Sensitivity to Intuition

While on the path of consciousness mastery, it can be helpful to live in a place where we feel joy and inspiration and are comfortable in natural wilderness. Being in surroundings of beauty and magnificence attract our attention to align with their vibrations. Walking barefoot on the Earth and swimming in natural pools and in the ocean can strengthen our alignment with the rising resonance of Gaia. Spending time in ancient forests and high mountains, where silence prevails, contribute to developing sensitivity to our inner knowing.

If we are not able to do this physically, we can do it in our imagination, although it can be a challenge to imagine absolute silence without experiencing it physically. For this, a trip into the high mountains on a beautiful day, or perhaps underground in a cave, is necessary. We can use inspiring music and deep, rhythmic breathing, bio-feedback, and other methods to help us be present in awareness. When we are present, observing from a perspective of positive, high-vibrations, we can respond to encounters, while we are in alignment with the energetic levels of gratitude, compassion and joy.

It's nearly impossible to develop inner sensitivity, while living in a state of stress or any kind of fear. Yet inner sensitivity is needed to live in a state of love and joy, because then we know that we have everything we need, and we know what to do in every moment. Making that transition from negative to positive polarity is a leap in consciousness, because of the need for sensitivity to our intuitive knowing. Once we make this leap, our life experiences change greatly, as long as we stay positive, which is how we are designed to be. This is what is natural for us, when we have no fear, only compassion and understanding.

True knowing comes from within our own consciousness. We receive it through our alignment with the conscious life force that we receive in every moment, and which connects us into universal consciousness. The clarity of our focus transmits what we know to our realization.

If we decide to engage in the exploration of our consciousness, looking for our limiting beliefs and examining them for validity, we can decide to resolve them in alignment with what we know intuitively. We can enforce our decisions throughout our innate consciousness with willful commands. We are the masters of our subconscious and can align our entire Being with positive, high-vibratory living in love, gratitude, joy and fulfillment.

Exploring Our Unknown Consciousness

Beyond our ego-consciousness, our unique personality, each of us exists as a magnificent Being, who is expressing our presence in a previously unknown realm of negative polarity. In our true infinite Being, we did not know polarity. Our human expression has given us the experience of duality, in positive and negative vibrations. Our experiences are the Creator's experiences. Our thoughts, feelings and actions are all enlivened by the Creator's conscious life force. What is unique about us, is our ability to direct the life force that enlivens us through our perspective, which we express as our thoughts, feelings and actions. We get to experience the vibratory level of our state of being, as do our Creator and all other conscious beings who focus on us. This is the reason that animals that we encounter in the wild know how we feel about them. When we align with our natural state of Being, we also can know how the animals feel about us.

We have learned to live within a defined spectrum of conscious vibrations, which we recognize as the empirical world that fills our perceptions. The animals are aligned with the consciousness of their entire species. We as humans also have a species consciousness. We all have a human perspective. We have similar feelings and ways of recognizing empirical energy patterns. We all have limiting beliefs that make it possible for us to experience life in duality.

The only ones who require us to experience duality as real, are ourselves. We can expand into a realm that is unknown to the human ego-compartmentalized consciousness. Beyond duality we are our present, timeless and limitless Self-awareness. Our true Being is available for our realization and alignment. If we can imagine the Creator living through us, experiencing everything we experience and feeling what we feel, we can become sensitive to the life-enhancing energetics that naturally are expressed in our life force. Our choice of the focus of our attention and our choice of energetic alignment determine the qualities of our experiences. We contribute to universal consciousness by our ability to modulate energy patterns that create experiences.

As we choose to open our awareness beyond the realm of duality in time and space, we open ourselves to universal consciousness through our intuitive knowing. Our self-identity transforms into a life-enhancing and light-filled person of expanded consciousness, aware of our presence in the consciousness of the Creator.

Working with the Masters

In our quest to raise our vibrations, while doing our best to keep our physical bodies vibrant, we can receive help from beings who have great wisdom and experience, as well as deepest love and compassion. They may or may not be embodied upon the Earth currently, but they have an energetic expression that we can interface with. We can imagine Jesus or any of the ascended masters or great angels being present with us. If we so desire, they will draw us into resonance with them. They can help raise our vibrations to the level of joy, ecstasy and bliss, because in our true essence, this is our natural state of being.

We share our true essence with the masters. We are all fractals of the One universal consciousness of the Creator of all, and we are participating in an experimental excursion in consciousness. We are experiencing life-diminishing energies that would be impossible for the Creator. The energetic level of human life, as we know it, is ultimately destructive and can only be created by our free will to enslave ourselves for this experience. It is the negatively polarized energies that we are contributing to universal consciousness, as it expands infinitely.

The great spiritual masters have learned that the entire world that humanity inhabits is an illusion in consciousness. We designed it to give ourselves real experiences, but it is a matrix of energetic patterns manifested by human life force through our attention and recognition mentally and emotionally. Although it is our conscious creation, It does not have the same reality that we have in our true Being, because it does not resonate with the Creator.

Because we are multi-dimensional Beings with the freedom to create whatever we want, we can live in more than one dimension. We can learn to control the quality of our lives as humans, regardless of anything happening around us. If we are motivated to focus on just being present in our awareness, feeling wonderful, and we imagine being in the presence of a great master, feeling the connection in joy and gratitude, we can raise our vibrations in alignment with theirs and begin to transform our lives into a higher dimension, where we have no interaction with negative situations and people. The energetic interactions and repulsions with our energy signatures make this possible. We can be fearless, loving and conscious of our eternal, present Self-awareness.

Aligning with Our Natural Life Force

Our role within universal consciousness is to make everything more harmonious, more beautiful, more interesting and more spectacular. We can imagine ourselves as super heroes, able to command more beauty, joy and majesty all around us. All of this happens within our own consciousness, and it is possible for us to realize this now. In our true nature, we are unlimited in our awareness and in our creative ability. Our ego consciousness cannot believe this, because we are programmed to be limited, but we can free ourselves from our limitations by our intuitive knowing.

Because we are fractals of Creator consciousness, we have been able to create the empirical world as a realm of duality and limit our awareness to it by creating and accepting limiting beliefs about our abilities and identity. Nothing keeps us limited, except ourselves. Once we realize this, we can examine our beliefs to determine their polarity and vibratory level. If they are negative, even just a little, they arise from fear and have no basis outside of our own perspective and energetic alignment. Every time we encounter what we believe to be a negative situation, we come face to face with our limitations. This is the time to confront them and resolve them.

Any energetic pattern that is negatively polarized has no creative sustenance in universal consciousness, which is beyond duality. The essence of universal consciousness is life-enhancing energetic expressions. For us in the realm of duality, that aligns with positive polarity, which is the nature of the life force enlivening us constantly. Negative energy is ultimately self-destructive and must be converted into other energetics. We have the ability to do this in our own consciousness.

By spending time in meditation in calming, natural places, we can train ourselves to become aware of our intuitive knowing in positive, high-vibratory visions and feelings. It's helpful to be grounded and In alignment with Gaia. By intentionally being in a state of joy and gratitude, we can receive our intuitive guidance more easily than if we are receiving dissonant vibrations. We can train ourselves to be in joy and gratitude as much as possible.

When we face challenges from negative energy, we can be in a state of love and compassion. The quality of energy that we face is within our own consciousness. If we are aware of it as such, we can recognize it as a sign from our intuition, asking for our attention in a higher perspective. In that state of being, we intuitively know how to think, speak and act. If we maintain a high perspective, we can resolve all limitations and realize our unlimited Being.

Consciously Entering the Void

Although we may know that we are constantly creating energetic patterns that we then experience, we may have difficulty being in the mental and emotional states that we desire. Our ego consciousness wants to plan ahead and evaluate events in our lives and react to everything we encounter. This is all non-productive for the spiritual path. We are designed to be in the moment and to be aligned with our intuition. Our intuitive guidance lets us know when to plan something. We just need to be open and present in clear awareness.

We do not need to wonder or be uncertain about anything. We just need to be confident in our inner knowing. This works best when we are just being present in awareness of the energy of the heart of our Being. In this state we can align ourselves with our conscious life force arising from the consciousness of the Creator, within which we have our Being. If we are open to it, it constantly enlivens and inspires us.

With our limiting beliefs about ourselves, we have restricted our access to our eternal essence and replaced our higher guidance with the ego consciousness of our limited being. For the ego, just being present in awareness is like abandoning control of our lives and possibly subjecting ourselves to disaster. It is this, actually, for the ego, because the ego has no ability to recognize higher guidance and must be transcended through our realization of joy and unconditional love in the level of energetics of our intuition.

Just being present in awareness is like going into the void of consciousness. As we practice this, our awareness expands deeper into our Being and out into the cosmos, which is a reflection of our own consciousness. We begin to realize universal consciousness, and we know everything we want to know. We can be in joy and gratitude as much as possible. These enable us to be able to be mentally and emotionally open, present and confident in our eternal Being. If we can maintain this state of being, we cannot be threatened or intimidated, because we are in a different polarity and vibratory dimension, where we are completely cared-for and supported in our creative ability. When we live in love and confidence, we have no fear or doubt, and life flows easily for us, with each moment bringing new awareness and creative potential.

4.

Aligning with Higher Consciousness

Aligning Our Entire Lives with Love

We experience the manifestation of energies that we have been in energetic alignment with. If we predominantly focus on being poor, we create the experience of poverty. The same with abundance. The only difference between the two is the vibratory focus of our attention. When we predominantly live in the vibrations of love, we enjoy mostly life-enhancing experiences. When we predominantly think and feel the vibrations of fear, we close ourselves off to some of our vitality.

We must raise our predominant vibratory frequency in order to experience more fulfilling interactions in all of our encounters. This means moving out of fear and into love. It's a leap in conscious perspective and may require a preview of our eternal essence of Being. We can ask for this awareness from our guides, angels and ancestors, and then begin to penetrate our deepest awareness.

When we open ourselves to the frequency of love, and we focus on going deeper into our awareness, we begin to feel and understand unconditional love. It is the energy that enlivens us and gives us our consciousness. This is our eternal life stream and our connection with the Source of our life in universal consciousness. It is high-vibration energy that our feelings can convey in deepest love and joy, when we are completely open to its level of vibration and come into alignment with it. Our energy signatures rise in frequency and radiate their energetic patterns into our enveloping, conscious quantum field for manifestation into our high-vibration experiences. Our lives become fulfilling in every way that we enjoy, and our interactions are compassionate, loving and supportive.

As we begin to feel the vibrations and to recognize our multi-dimensional greater Self, we can recognize the compartmentalized level of consciousness that humanity lives in. We are not required to align ourselves with this level of consciousness. It is a free-will choice that we made prior to incarnation, which required becoming unaware of our full consciousness. We have the ability to regain our full consciousness from within our limitations. It requires intentional, deep penetration into our conscious awareness, and learning to recognize, feel and align with higher frequency patterns of energy. If we desire to live in the energetic spectrum of love, joy and abundance, we can imagine ourselves experiencing these vibrations in any scenarios. In the course of our lives, if we can stay predominantly in the vibratory level of love and compassion, our experiences come into alignment with our thoughts and feelings.

Continuing Conscious Expansion

We have been held back from realizing our greater Self beyond

Chapter 4. Aligning with Higher Consciousness

time and space, because we have been afraid of "what if" disempowering situations. To make the leap into the awareness of love everywhere, we must leave all fear behind. There are no "what-if" situations, unless we, on some level, align with fear of them. Without fear, we can open ourselves to miracles, which are just excursions into a higher dimension, and which then manifest their vibrational level in our experiences. Our lives can become magical, once we learn to control our attention and maintain high emotional vibrations.

The secret of expanding consciousness is love. Fear contracts and shuts down our flow of life force. Love expands our life force and our radiant presence. Love enables us to know the source of our life force and fills us with gratitude. We're uncomfortable around fear, because it's not our natural vibration. Fear is a spectrum of vibrations that we could not experience, unless we closed ourselves off to our Self-realization. As we become aware of our true Being, fear fades away, for it needs our life force to exist. We've experienced what we needed in order to develop deep compassionate wisdom.

We can now return to our unlimited consciousness and abilities, our true Self. We can be comfortable in the presence of love. This vibratory level leads to an awareness of universal consciousness. We can know the essence of our Being and realize that the same essence envelopes us in every pattern of energy and everything around us in the quantum field. We are consciously personal participants in the field of energetic vibrations, which we can recognize and transform with our imagination and emotions into high vibratory experiences. We are multi-dimensional Beings, and we can open ourselves beyond the boundaries of dimensions. In the spectrum of the energy of the heart of our Being, we can realize our eternal Self. Forever after, fear is only a memory.

We can live in whatever vibratory level we choose, because we are the Creators. We are the personal aspects of the Universal Creator consciousness, from whose Being we arise to provide experiences that enrich the universal consciousness. In our current human dimension, we've provided all manner of low-vibrational experiences of fear. These have deepened and expanded our conscious awareness beyond our natural state of Being. As we leave all of this behind, we can envision and feel for more beauty, freedom and love. We can free ourselves from the limitations we've placed upon our awareness by imagining living in wonderful experiences. By living consciously beyond time and space, we can become masters of the current spectrum of human energy.

Aligning with the Heart of Our Being

We can recognize our heart as the physical source of our life. We can also recognize the heart in its expanded, higher-dimensional, essence, as the receiving and distribution point of our conscious personal life force. Our heart lives only for our well-being and vitality, regardless of what we do to it. The consciousness of our heart is the source of our ability to love and be compassionate. These are the feelings that emanate from the heart of our Being in universal consciousness.

We can search for the most uplifting feelings that we can imagine, and we can keep expanding upwards emotionally, until we allow ourselves to be enveloped by the unconditional love and joy of universal consciousness. This is the natural state of our true Self. We are the Creator, manifesting as our personal Self, whom we have erased from our awareness in order to live in the compartmentalized consciousness of humanity's resonant frequency.

The empirical world entranced us into not even wanting to expand our awareness, but we are being prompted to do so. As the resonant frequencies of the energy of the Earth rise, things in the lower frequencies become chaotic and threatening. There is another choice for our attention. We can choose to

Chapter 4. Aligning with Higher Consciousness

pay attention to the energy of our heart and the emotions that it expresses.

Our heart loves us unconditionally, regardless of anything. We can search for the feelings of being everyone we encounter. Our biophotons interact when we come into each other's aura. We can feel each other's presence, and we can know immediately each other's vibratory level. We can see the light in our eyes. We can choose to feel connected in consciousness through our energetic interaction. Our heart consciousness urges us to live with high-vibration emotions. All resonant vibrations that enhance vitality and stimulate creativity arise from our heart. They are the feelings stimulated by the energy of our true Self flowing through our heart.

Our heart offers us higher guidance beyond our ego consciousness. It constantly prompts us through our feelings and intuition with what we need to know in every moment. If we have disregarded this guidance for most of our life, our inner guidance is still present. We can search ourselves for it. It includes our conscience, and is much more. We can only know it by experiencing it. It is our inner knowing. It's how we know, apart from empirical proof.

Moving into Eternal Freedom

There is a galactic-wide rising consciousness occurring, and the Spirit of the Earth is expanding in consciousness, causing the frequency spectrum of our planet to expand and rise in vibrations. This is the direction of our life stream. We are destined to expand our awareness into higher vibrations of living.

Throughout history we have been tricked into allowing ourselves to be enslaved. We are naturally free Beings and cannot be enslaved, but our compartmentalized human consciousness does not know this. Energetically the issues involved here require a leap in consciousness. We must be willing to be the best person we can imagine ourselves to be, regardless of results. This is a leap beyond ego-consciousness and requires higher vibrations of mental and emotional states of being, regardless of what we are encountering. We must be willing to be finished with suffering, pain and fear of termination. We must change our emotional polarity to positive, high-vibrations at all times, knowing that we are cared-for by the structure and operation of the cosmos. By living in high vibratory energy in our thoughts and feelings, we attract that kind of energy into our experience. With this perspective, we are available to our intuitive knowing of how to interact and live.

When we have emotionally freed ourselves from fear, threatening situations can transform into positive energy, or they disappear from our experience. We can be confident and grateful for our natural well-being and abundant level of living in freedom and sovereignty, not wanting anything from anyone. We can be completely Self-sufficient. Our intuition guides us constantly in everything we need to know. We only need to pay attention to it and follow its guidance in the vibratory level of love and life-enhancing thoughts and feelings. We can be vivacious and filled with vitality, because only good things can happen in our experience. This is the world we are creating from the energy of our heart.

Working Harmoniously with Our Emotions

Our emotions are a mysterious part of us, because we're so accustomed to understanding through our mind. Emotional intelligence, however, is more powerful in guiding us through life. When we become overwhelmed by emotion, our mind has little ability to do anything. Our emotions immediately know the quality of energy we are encountering in every situation. They also radiate our state of being to all around us. They are intimately connected with our body, which has its own physical intelligence. They are the source of artistic expression and social interaction and cohesion. They are the motivating force for all of our actions and beliefs.

We can expand our consciousness only with the help of our emotions. We have trained them to support our ego consciousness and to function in the energetic spectrum of the negative polarity and low vibrations of fear and life-diminishing feelings. This has resulted in depression, shame, lack of trust and closing ourselves off to true love. Our world appears to be full of threats to our well-being, and our emotions respond to these, regardless of what we may think. This is why many artistically sensitive people have dramatically difficult lives.

When we're in the grip of intense negative emotion, how can we transform it? We can calm ourselves through deep, rhythmic breathing. The breath sooths our emotions and our mind and unbinds the energy of our heart. It can bring us to a point of neutrality. We can elicit higher vibrations from our heart center, once we are emotionally serene, and we can transform a low-vibration experience into feelings of compassion and kindness.

If we are searching for our true Being, we can become filled with gratitude for our living awareness. If we can know this, we are expanding our awareness. From here it is a short step to knowing that our awareness is eternal, because we're expanding beyond the body. We can become aware of things we cannot see and could otherwise not know. Progress from here depends on our trust of our inner knowing. It is in every way far more comprehensive than our rational knowing.

At first we can intend to be sensitive to our inner knowing. We can learn to pay close attention to our promptings and act in alignment with the energy of the promptings. This takes us into higher visionary vibrations. We can be guided into a wonderful life, full of love and joy. This is our natural state of being. It is what we are created to experience, and it is available to us whenever we awaken enough for a meaningful intention.

The Story of Our Lives

We know about electricity, and we know about magnetism. They are both energetic waves that travel together within the quantum field in planes at right angles to each other. At the point where they cross each other they can manifest in our experience, if we recognize them. This is how we create the material world. In our own being, we have the ability to change energetic waves in the quantum field, which we dwell within, into experiences in our senses by becoming aware of them. There are infinite patterns of electromagnetic waves for us to experience, and we have the free will to choose which ones we want to recognize.

We have become accustomed to the vibrations within the spectrum of energy that humanity dwells within, but there is so much more that we are capable of knowing. Because we have compartmentalized a portion of our consciousness to be able to experience energies that we would never encounter in our true Being, we have been able to live and experience the low frequencies of fear and evil. We have done this in order to deepen our sense of compassion and love. We've designed this compartment of consciousness to be so entrancing and empirical that we could not escape into the higher vibrations of our true Being without a strong intention. We are being challenged to do so from within this compartmentalized awareness.

We have realized and created ourselves into our predicament, and now we must create ourselves out of it. We cannot do so by our actions. We must do it with our consciousness. If we focus upon our problems and enslavement, trying to resist and change things, we send our life force into the vibratory level that we do not want, and we just continue creating low-vibration situations for ourselves. We can resist by refusing to participate, but unless we raise our own vibrations, we are still subject to invasive, low-frequency energy.

The only real way out is to focus on the energy of the heart of our Being, symbolized by our physical heart, which embodies the unconditional love of our Creator. Our heart lives only to enliven us and give us vitality, regardless of what we do to it. This is the high-vibrational energy that we must align with in order to know our true expanded consciousness. Focusing upon the deepest love we can imagine and feel, along with personal clarity of Being, we can begin to rise in vibrational frequency and know the truth of who we are. At this level of feeling and envisioning, we rise above all low-vibrational experience and cannot be threatened, because we can know that our personal awareness is eternal, free and sovereign. We can know this by intentionally being in continuing, life-enhancing, high-vibrational feelings and thoughts that bring us great joy.

Learning How to Cope with Humanity

The only thing that matters at the end of our human life is love. On the deepest level it is the energy of who we truly are. When we celebrate with family gatherings for special holidays, we raise our vibrations in love for one another, and we help those who could not celebrate. So it is in the higher dimensions. Everyone can participate in high-vibrational living with much joy and abundance. It is natural for us to want to help those in need.

Many of us feel that we don't need etheric help. If we continue with this perspective, we will not receive it, even though it is always available. Our human perspective, based upon ego sustenance, can operate only in a negative polarity, low vibration environment. If we can change our polarity to high-vibrational love, we enter a higher spectrum of energy, and all of our experiences are elevated. We have freedom of choice in everything and are sovereign and free in our awareness. Our conscious awareness never disappears, as is reported by everyone who has had significant out-of-body experiences. Our awareness actually expands tremendously when we're not focused within the empirical world.

Since our human presence is limited by the thoughts and feelings involved with our personal beliefs, we cannot imagine much experience beyond the empirical world. Even the concept of God has human attributes. The quantum world is mysterious to a person with an empirical perspective. It is a realm of electromagnetic wave plasma and patterns of waves of every possible eventuality. It envelopes us and inter-penetrates us. We recognize energetic patterns that we align with in our perceptual and emotional ability. These become our empirical world. They are oppositely polarized and of a lower vibratory frequency than our natural state of Being.

The empirical world has a denseness that is not present in the higher realms. Part of our desire to live in human bodies of this density is to feel the pleasures of the senses. What has happened in this density is that we have not allowed ourselves access to our higher guidance. We have filtered it out. We live in a realm of cognition controlled by the ego that we designed for this purpose, as required for our human experience. It has become a limitation that cannot change without changing its polarity, causing a complete change of identity.

Once we intend to recognize the importance and essence of the love of our Creator Being, we are on the path to awakening to our inner guidance, because that is the vibratory level of our intuition. It is part of our innate Being, which controls all of the life processes in our bodies, knows us better than our self-awareness, knows everything that we have ever experienced in all of our lives, including all of our thoughts and emotions. It has only compassion and unconditional love as the essence of its Being. It is the consciousness that provides the experiences that align with the levels of vibrations that we dwell within predominantly. It gives us experiences within the spectrum that we choose to vibrate at.

We are in charge of our innate Being. It reads our vibrations and acts to accommodate us without judgment, only love and compassion. When we suffer, it is because we have aligned our vibrations with those that manifest as suffering. When we give our focus and life force to enslavement, as has been the case with most of humanity, whether through submission or resistance, we align with the energetic level of slavery, and our Innate manifests it for us.

As we intend to go higher in our joy, we become more expansive in our awareness. In order for this to happen, we must engage our innate Being. We can address our Innate, and tell ourselves how we intend to be, and thereafter live on this vibrational level. Our innate Being can cure any ills in our bodies instantaneously, when we are in complete alignment with the energetic expression of vitality and well-being. And so it is with everything. Whatever we imagine and align with emotionally results in experiences that are compatible with that energy. We believe and know ourselves into our experiences.

In order to change an intense experience within negative, low-vibrations, we must extract our awareness from that scenario in order to be serene and reorient ourselves. Deep, rhythmic breathing for a while and perhaps a walk in nature can do it. Training ourselves to believe what we want to experience is necessary. We have to be able to recognize and feel it, to recognize and feel the vibratory pattern.

As we become proficient in envisioning and, in our awareness, being in high levels of vibrations, our experiences come to reflect these energetic expressions. We can choose to live in the vibratory level of gratitude, love, joy, abundance and freedom. We can choose to know our eternal, unlimited awareness, which we share in the universal consciousness of the Creator.

Understanding Jesus

Jesus told us that we could do everything that he did and more. Either no one believes that, or we have not properly understood what he meant. Let's look from his perspective. He was a Self-Realized person. He knew his eternal essence of Being and his multi-dimensionality. He lived constantly in high-vibratory frequencies of emotions and thoughts, even when confronted with powerful threats to his humanness. In that level of Being, there can be no sacrifice, because he was the master of all reality and in full control of his unlimited creative ability. He said that we are like him. He meant that we are all fractals of the Prime Creator. The difference for us is that he knew all this and could modulate the energies of the empirical world. By definition we have the same abilities, except that we limit our awareness of them.

Jesus vibrated as a human being in a higher dimension of frequencies. He established that frequency level for us. We are of the same essence of Being. We have eternal presence of awareness. Our conscious awareness cannot be terminated. If we mostly hold our attention toward higher vibrations in each moment, we will keep expanding through our limitations. Intuitively we know that we are eternal. This is what having a positive electromagnetic polarity with high-frequency energy patterns offers us. In the current world of humanity, nearly everyone has a negative polarity. Humans believe in termination of conscious being.

Making the shift to the positive polarity of love and joy is a leap in consciousness. Yet this is what our intuition is directing us to do. We can feel it and know it, if we want to. We can drift back and forth, but once we truly experience high-vibration living, we can no longer be seriously engaged in the drama of human life.

We can be the masters of every situation by understanding our connection with universal consciousness. We can be aware of many different arrangements of circumstances in each moment and can choose the one we want to experience. This may be what Jesus meant. The Bible just leaves out the part about Jesus' level of conscious awareness, but we have the stories that show that he knew how to command the elements, as can we, when we are at his polarity and frequency levels.

Aligning with the Infinite One

Most of us here continue to believe that we are finite and mortal. We live as if we are. These beliefs are our subconscious programs. Through intentional training, we can repolarize these programs and resolve them. One way of doing this is to make the conscious leap to positive, loving thoughts and feelings. This perspective can result from our focus upon our present Self-awareness. We are eternally present awareness. We can engage joyfully with others, as we become aware of our infinite, unlimited creative ability and power. This is our natural level of consciousness, and it is available to us. It feels really good.

In every moment we make a conscious or subconscious choice of the level of vibrations that we create. Every thought-feeling creates the quality of our experiences. We can be aware of our choice of the quality of energy that we focus upon in each moment. By constantly choosing to imagine and feel high-frequency vibratory patterns of compassion and understanding, we can resolve our limitations. We needed them for our human experience, but once we really trust ourselves to be true in living for the good of all around us, we no longer need limits. Through the conscious life force that eternally flows into our Being, we can connect with the unlimited consciousness of the Creator.

Because we are fractals of the Creator, we have every possible capability. But we haven't believed this, and this belief has kept us from destroying the cosmos. We were learning about negative energies and everything that is life-diminishing. We needed a safeguard, in case we went too deep into low-vibrations. When we have learned that we prefer positive energies, we are ready to leave the lower dimension and make a conscious leap into the higher state of Being. Here is where we can regain our powers and align ourselves with the conscious vibrations expressed by the Creator.

Our intuitive knowing is our connection with higher consciousness. It has been guiding us since birth, but its guidance becomes faint, if we fail to give it our attention. The role of our social training has been to distract us from our inner knowing. Being sensitive to our intuition is the most important thing we can do to align ourselves with higher guidance. Our guides and angels can also influence us this way. We can practice and learn to be present in awareness in every moment and open to changes in everything.

We can learn to trust our creative abilities and use them for wonderful energy modulation. Our visionary and loving imaginary scenarios can vibrate at high frequencies, raising our personal energy signature's vibrations and expanding our awareness. We can do this with every situation we find ourselves in. By being always positive, we open our experiences to the energetic patterns of joy and abundance.

Reaching for Fulfillment

As we expand our awareness, we become brighter and more ethericaly radiant. We can imagine being in the presence of luminous Beings, interacting with them and aligning with their vibrations. It can be an experience of great beauty, pleasant fragrances and music and wonderful, clear brightness and love. We can meditate on the Sun and align with its luminosity. Imagine a Being so incredibly radiant that it warms our planet at great distance. Eventually we all become brilliant, while we are returning to the consciousness of the Creator.

We are beings of light. Each of us is surrounded by an etheric aura of photons of many colors. As we grow in awareness and love, we emit more photons, making us more radiant. As we become clear and attuned to our intuitive knowing, our aura begins to become visible. We are becoming our true Selves, and we are able to live by higher guidance through our intuition, without any personal drama. We still have our roles to play, but we can transform any low-frequency encounters with our energetic presence. We can maintain a perspective of compassion and understanding.

Once we know that we are sovereign Beings, we know that we cannot be threatened, because we are also eternal in our Self-aware consciousness. In this state of Being, we need nothing from anyone, because we know that we always live in abundance. Our thoughts, feelings, speech, and actions are for the benefit of everyone in our awareness, including all of nature, our planet and beyond.

We know what energies are life-enhancing, and these are the ones we can choose to focus upon. This does not allow negative, life-diminishing energies into our presence and awareness, unless they are presented to us for transformation through our compassionate wisdom and higher guidance.

It is no different for anyone, except for our state of being in every moment. It is our polarity and level of vibrations in our thoughts and emotions that attract and repel people and experiences. These also cause scenarios to develop. We can choose to guide and create energy patterns that express goodness and fulfillment in each moment.

We Can Intensify the Rising Energetics of Gaia

The Spirit of the Earth has had a predominant resonant energetic frequency of 7.83 cycles per second. This is very slow compared to the frequencies of our technological communications, but it is the frequency of life on this planet. We have all aligned with this frequency in order to live here. Now, however, due to external cosmic-ray injections, as well as changes in the consciousness of Gaia, her resonant frequencies are showing increasing levels of energetic quality in several octaves higher. To this development, we are all having to adjust and evolve spiritually. The Earth is going through a physical and spiritual metamorphosis.

The effect of this development upon humanity is becoming

dramatic. It is becoming more and more obvious that the negative, low-vibration evil is coming out of the shadows, and those aligned with it are becoming uncomfortable being here. Those who were the most negative have become so unstable, that they have already left. Those who remain have become disoriented and delusional. This includes most politicians and corporate officers. The rising frequencies are coming into closer alignment with the conscious expressions of the Creator. This situation is affecting all conscious beings.

There are cycles in consciousness at very long wavelengths. Some are longer than we can measure or calculate. We were at the bottom of a 25,920-year cycle, and we're on our way up. The decision of the consciousness of humanity as a race of beings, to deepen our understanding and experience within a conceptual range of frequencies that we would otherwise never be attracted to engage with, has compartmentalized our awareness within the negative polarity, fear-based frequencies. If we can change our polarity and raise our vibrations, we can move into greater alignment with higher consciousness and move into the vibratory spectrum of love and its compatible frequency patterns.

To change polarity means to live mostly with positive emotions and life-enhancing thoughts. This can be an intimidating change. It transforms our personality. We may change friends and even partners, who no longer align with us. We may change nearly every aspect of our lives. Our process is like an MC Escher graphic design, in which everything, which is a fractal of the Whole, changes mathematically (In this case, us) continuously in complete synchroneity with the One universal consciousness.

From the negative perspective, this is all unimaginable as a viable lifestyle. There have to be miraculous shifts at times in order for things to connect properly. How can this happen? It is because consciousness is the creator of every situation, and the quality of any situation for us depends upon our own resonant frequency in comparison with Gaia. Our alignment adds our own conscious life force to the ascension energies of our planet.

Expanding Consciousness in Gratitude, Joy and Compassion

In our Self-aware life stream constantly flowing into us out of the One conscious Being, we have the potential of knowing infinite awareness and creative power. When we seek to recognize our higher Being, our awareness begins to expand greatly. We can feel the previously unrealized higher-vibratory emotions that come to us in our life force. As we intentionally align with the vibratory level of gratitude, joy and compassion in every moment, we can access our greater consciousness beyond time and space, and we can become our eternal Self-realized presence of awareness.

The greater we can open our awareness to higher vibrations of every kind, the more unlimited we become. We become expansive in our expressions through our imagination and emotions. We can choose to live within the positive polarity of unconditional love as much as we can imagine it and feel it. As we come into closer alignment with our intuitive knowing, we can transcend any limitations. We can understand them from a more expanded perspective of greater wisdom and compassion. By our high-vibratory perspective, we attract situations that align with our energy signature. From the perspective of the ego, things become magical.

As we continue to desire to open to universal consciousness, we begin to recognize the essence of all Being. This recognition allows us to enter the consciousness of all beings. We can realize the awareness of our pets and favorite plants. We can imagine the presence of nature spirits and become aware of their awareness. We can elevate the presence of everyone whose awareness we enter through our own energetic radiance.

We were created to enjoy and celebrate life in all ways. Even now, we have that potential. We can have so much fun with our consciousness, once we free ourselves from our limiting fixations and open ourselves to eternal awareness.

The Greatest Life-Enhancing Energy

In our traditional human consciousness, we know the words of unconditional love, but we do not know what it feels like or actually is, because it is beyond the conception of our ego. It is a state of consciousness that is the greatest life-enhancing energetic expression possible. It includes the consciousness of every being in all dimensions. It is the energy that we constantly receive in the flow of conscious life force within universal consciousness. By being in alignment with unconditional love, we become aware of our own awareness and the awareness of others, all the way down to sub-atomic waves/particles and out to the universes. It is all enveloped within One consciousness, which we naturally can participate in. We are of the Consciousness of the Creator of all.

Our bodies are manifestations of our consciousness. To the extent that we have ailments, we have aligned our vibrations with beliefs and personal drama that create anomalies in the flow of life force through our innate consciousness, resulting in life-diminishing developments. Our deepest beliefs and fears have stayed with us through many lifetimes and can be recognized and resolved in this one.

By imagining and feeling, as much as possible, that we are living in a high-vibration life-style of great joy and freedom, we can open ourselves to great improvements in our lives. According to quantum physicists, our entire experience as humans is created, dissolved and recreated trillions of times every second. Because we constantly create the quality of our experiences by our state of consciousness, we can change everything in our experience by intentionally changing our conscious awareness.

Our random thoughts are not our own. They are the electromagnetic wave patterns that we recognize in the quantum field of all possible waves and patterns of waves. The patterns that we recognize with our mind and focus upon are our thoughts. We have absolute freedom to think and feel however we choose. In terms of our state of consciousness, it is the polarity and vibratory frequency of the thoughts and feelings that we choose to focus upon that are constantly created in our experiences. We can momentarily create and change the polarity and vibratory level of our awareness. This creates alignment with compatible scenarios for us to experience.

When we have no personal interests or attachments, we can become aware of our eternal presence of awareness. Then we can realize that we have lived as humans in a compartment that is closed-off in our consciousness. Everything that happens in this compartment is designed to distract us from Self-Realization. It is all beginning to fade away, as more of us realize our situation and begin to focus on a higher dimension of energetic vibrations and experiences. We can begin to enter the conscious awareness of the Creator.

Realizing Greater Truth about Our Being

On the inner path to divine consciousness, we can begin to expand into unlimited awareness. If we can open ourselves beyond our personal attachments and beliefs, by being in a state of deepest love and joy, we can realize our greater Self. In our true Self we are our eternal presence of awareness with unlimited capabilities of creating and manifesting. This is the state of Creator consciousness, of just Being in unconditional love and universal consciousness. It is Being everyone and everything in the shared consciousness of us all.

It is recognizing the game that we have been playing as humans. We have been very convincing actors, so good at it, that we even believed that we are the characters we've been playing. We created this game to be as realistic as possible, and we really succeeded. Now it's ending, and we have the opportunity to expand into a higher dimension of living, a dimension of positive polarity and higher vibrations. There's more joyous living.

This is a higher dimension of our game of living with stimulation of our senses, emotions and mind. Once we learn to follow our intuitive knowing, our mind has a very limited role.

We can know the consciousness of the Creator and align with it through following our intuition in every prompt we receive constantly. It's like a higher dimension of learning how to ride or drive. It takes practice in a new kind of sensitivity, and then it becomes automatic.

This is how we can train our limited awareness as humans to open up to greater possibilities. If we are willing to live in an imaginary wonderful realm, and we are willing to train ourselves to recognize this realm, it comes into our experience when we believe it's real. This happens just as we recognize it, and It can happen in any dimension, according to our polarity and vibration. We can challenge ourselves as much as we want and more. Some of us have become incurably stuck in dark energy, but we all have absolute freedom to choose any and every kind of energy in our awareness. Our state of being expresses the level of vibration and the quality of life that we create in our experience. Forms arrange themselves for us, according to our state of being.

Being Aware of Being Aware

Because consciousness is universal, we can be aware of the awareness of any conscious being. We live in a sea of consciousness with an unlimited number of energetic expressions that are constantly expanding. We have the ability to be aware of any of them, but as humans, we have limited ourselves to the empirical spectrum. We also have the ability to unlimit ourselves. We can do this by realizing the feeling of negative polarity experiences, while staying in a higher state of Being. The negative experiences cannot exist without our life force, which we send out with our engagement. By transcending engagement through compassion and wisdom, we can resolve our self-imposed limiting beliefs. We can free ourselves to become our eternal, unlimited Selves.

As electromagnetic beings, we maintain a state of consciousness that is largely dependent upon our emotional state. Emotions operate magnetically at the vibratory frequency that we feel. Our thoughts are modulated electrically, as are all of our neurological systems. They cannot operate independently of our emotions, and must come into alignment with how we feel.

In searching for our intuitive guidance, we need to be in a calm, stress-free emotional state. This we can find in nature, where we can align with the energies of Gaia and the spirits of nature. Connecting with the vibrations of the Earth is most important. We can also listen to soothing music, and we can meditate and practice tai chi and other centering kinds of movements. We can watch inspiring videos. Whatever inner path we choose, practice creates the ability to calm ourselves easily, even in chaotic situations.

As we learn to be in a state of emotional neutrality, we can gain the ability to be aware of the awareness of our ego consciousness. We can observe our state of being in every moment and bring ourselves into alignment with higher vibrations in the spectrum of love and compassion. This is where we can be aware of our inner knowing of the constant guidance that comes from the consciousness of the Creator. In the face of the synthetically-created chaotic energy of our time, we truly need the connection with our intuitive knowing in the energy of our heart. It is our vibration that is elevating humanity. We are creating this high vibration through our awareness of our greater Being.

Realizing Our Inner Light

Although we have potentially unlimited abilities, we have been taught and programmed to believe that we are physical humans with abilities limited by our empirical presence. Because we believe in our limitations, we create them in our experiences. We can, however, intentionally open ourselves to our higher knowing. By asking for inspiration from higher-vibratory beings, and searching for higher guidance in ourselves, we can realize our true nature.

Our power center is our heart. It emits the energy that powers our body, and it is the power center of our conscious life force. It gives us Self-Realization, which we can know, once we drop our self-imposed limiting beliefs and open ourselves to being unlimited and fully in control of our lives. Once we open ourselves to our intuitive knowing, we no longer need beliefs that limit us, because we can trust ourselves to be true.

We can begin to realize that we are becoming brighter in every way. Our heart emits unlimited numbers of photons, which are in our visible light spectrum, but below the threshold of our human sight. Because of its etheric presence, those who have opened their pineal gland and have etheric vision can see the aura. If we attune ourselves to the intersection of sound and light at 528 cycles per second, we can hear the music of our aura.

Through the etheric energy of the heart, our intuitive knowing comes to us. When we open our awareness to it, it fills our being and offers great vitality. It draws us into alignment with the consciousness of joyful positive polarity. On every level, our heart enlivens us, regardless of how we choose to treat ourselves. It is the source of our brightness, our inner light, which we can learn to see with inner sight, and we can feel it, even in ordinary human awareness.

We can feel our own presence of Being, as well as anyone else's in our presence. We can immediately know the quality of vibrations of anyone, including ourselves. Our intuition tells us through our emotions, even if we're not consciously aware of it. We acquire this awareness through our intentional openness to it.

We do not need to be captive to negative, low-frequency energies, as most humans have been. We have the ability to free ourselves by intentionally being positive and seeking to live in the high vibrations of joy, compassion and gratitude. We can align with the energetic patterns of Creator consciousness by feeling what these are and by searching for the music of our inner light at 528 Hz. Once we find this, it can transform our physical DNA and carry us into the unconditional love vibration of Creator consciousness. It is a perfect note for OM chanting. Check it out at https://www.youtube.com/watch?v=9PRV6w6VJbc.

Knowing Intuitively

Our ego-conscious mind does not know what intuition is. It is in a different spectrum of energetics. The closest that the ego can come into contact with intuition is through higher mathematics. Mathematicians reach beyond the capabilities of the ego mind into pure logic and intuitive solutions. They, however, do not know how their solutions come into their awareness, except that they intend to know, and that is the creative link.

Intending to know our intuition, and being attentive to what we deeply feel and know, can deepen our sense of intuitive guidance. This guidance is always life-enhancing. It brings us into alignment with our true Being. We can imagine being our divine Presence, and we can intuitively know our divine, eternal Presence, if we want to. In this state of Being, we know whatever we want to know. We can imagine and feel the quality of being of everyone we recognize in unity with the essence of our Being.

If we keep living in this knowing of the quality of all the energetic patterns available for our recognition, we can choose to focus on expressions of love and goodness, compassion and joy. These are the positive, high-vibratory feelings of our higher-conscious Selves, and they accompany our intuitive knowing. There are no inherent limits to our intuition, other than our beliefs about ourselves.

Our beliefs keep us hypnotized in the dimension of human

experience. In this realm our awareness comes only through our ego-consciousness, and we are not capable of awakening, unless we want to. Once we intend to open ourselves to greater realization, situations arrange themselves to accommodate us. If we continue intending to expand our awareness in love and compassion, our intuition is our constant guide for insights and knowing everything we want to know. We can become sensitive within to receive guidance from the universal consciousness of the Creator.

Resistance or Acceptance of Evil

Whether we realize it or not, we are all moving into the light. It is the positive, high-vibratory direction of Gaia, and all who live here are living in her rising vibratory resonance. Soon there will be no negative polarity, because Gaia is only positive in her rising life force. We are in the throes of growing awareness of the most evil energy, as it becomes exposed to the light. The power of its polarity is diminishing, as it shuts itself off from the divine life force that it needs to exist. Its existence must depend upon the alignment that humanity provides. All of the evil ones are parasites who steal human life force. We can resist them and feed them our energy, or we can love them into transformation or dissolution.

The energetics of our situation can provide either polarity. Resistance to government tyranny carries us into alignment with that level of mental and emotional vibration. We give it our life force, so that it can exist. It could not exist without our support. Instead of engaging negative energy with negative energy, even if it seems right, we have the choice of being in a higher state of awareness. By maintaining a positive, loving and compassionate perspective, we cannot be intimidated by the negative. We are the creators. We can create experiences that are in harmony with our level of conscious love and joy.

We can recognize the feelings and images of evil in those who are in alignment with it, and we can understand them with compassion, knowing that they are playing roles that we've been involved in during some lifetime. That's what we've been here to experience, to know intimately how evil feels and how averse it is to our own Being. By resisting it, we align with it and give it our life force through our attention and emotional strength. By accepting it with compassion, we can be in a loving state of being, and we can open the way to living in a higher dimension.

In a state of positive, high-vibratory energy, we can create the new world of life-enhancing energies. We can be in this state of being all the time, regardless of external challenges. With the guidance of our intuition, we know what to do in any encounter. Our intuition always has the level of vibration that streams through the heart of our Being. It is a stream of vitality and unconditional love in universal consciousness. This is what we can align with and transform our experiences.

Designing and Creating our Reality

The great force of life that fills the cosmos expresses Itself through us. We are the energy modulators for the creation of everything we experience. Through our conscious perspective, our mental and emotional energetic alignment, we feed our life force to whatever entity and energetic pattern that we engage with. It is a simple process. By imagining ourselves as separate from those we consider evil, we create separation and enmity. By imagining ourselves in unity with all and aligning ourselves with unconditional love, we create a world of kindness and compassion.

This may seem impossible to our ego consciousness, and our experience in the world may consider this to be foolishness. How can we not resist and punish those who diminish and destroy our empirical lives and expect that anything good could come of our interactions? It is a matter of understanding how energy works and what Jesus meant when he instructed us to turn the other cheek.

In resisting evil, we must engage it on its own level of consciousness, its polarity and vibratory level. We cannot defeat it there, because it is the natural expression of that level of consciousness. We can, however, transform it in our experience by accepting it with compassionate wisdom and asking to know the highest-conscious response within our innermost Being and the energy of our heart. Here we are in a natural state of joy and gratitude for our entire Being. Circumstances will arrange themselves to accommodate our state of Being. Our potential is unlimited, creative Being.

We can no longer be completely limited to the energetic spectrum of traditional humanity. It happens in our alignment with life-enhancing energies, those that have a positive polarity and high vibrations. These are the natural energies of our true Being in our expanded state. They are the feelings of love, gratitude and joy. We are capable of being in this level of consciousness through our intention and devotion to truth on all levels.

The negative energies will just fade into another dimension. This is the veil that humanity is awakening through. As more of us withdraw our life force from negative, low-vibratory thoughts and feelings, we can focus on the highest energies we can engage and align with. This is an individual and unique process for each of us. We can do it with our thorough and true intention. We know all of this deep within. This attracts compatible energetic patterns that provide the quality of our experiences.

Transcending Ego-Consciousness

We are always in the presence of our highest Being. To be in joy is our natural way of feeling. It is what we feel, when we open our awareness to it and realize it. Joy is an eternal expression of life-enhancement. By intentionally feeling ourselves in joyful situations, regardless of whether they are physically threatening, we can be aware of our intuitive higher guidance in every moment.

Our ego consciousness is deeply set in our compartmentalized consciousness as humans and is negatively polarized with fear. Changing our perspective to positive is not possible for the ego. It requires a leap in consciousness that can happen with strong intent in transcending the ego. If we can make the leap to compassion, gratitude and joy in every moment, our experiences arise in this level of energy, and life becomes miraculous to the ego.

Transcending the ego can happen by just imagining being joyful and thankful as much as possible. It means refocusing from feeling threatened by anything to being grateful for every encounter, because each is given to us as we created it with our constant creative ability in our thoughts and emotions. We live in the spectrum of vibrations that we create, and we attract situations that vibrate in resonance with us.

We can elevate our lives by intentionally imagining and feeling that we are living in loving and joyful environments within and without. By being positive and open, we can expand our awareness to more wonderful possibilities for us. As we radiate positive, high-vibratory energy, we also attract it, and our lives come into alignment with higher consciousness through our intuitive knowing.

Once we realize that we are our own personal eternal presence of awareness, unlimited in every way, we can understand our human self in the roles that we chose to play in the negative experiential realm. We desired to deepen our understanding and compassion in this realm. In our true Selves, we can be masters of the material world, able to create lives that radiate joy and love for all. We can recognize the light of the life force that flows into all conscious beings and interface with it, instead of facing egos.

Adjusting Our Perspective

Once we are in alignment with the life-enhancing energy patterns of the heart of our Being, we can be aware of our life experiences from the perspective of greater awareness of our creative ability. We have had no idea of our abilities beyond the energy spectrum of humanity; however, we can play with our awareness in conjunction with our imagination and emotions. We can pretend to be our higher Selves, unlimited in power and creative ability. This is one way of moving beyond our personal limitations.

Our innate awareness is unlimited in all dimensions in which we choose to participate. We always have choices for the focus of our attention. We can choose our mental and emotional polarity and our vibratory level. Eventually we can open our realization to universal consciousness. Our process depends entirely upon our polarity and vibratory resonance. These are the qualities of all energy patterns, and they stimulate our thoughts and emotions.

From a perspective of heart-consciousness, infinite love and compassion, we can recognize the qualities of energy around us and throughout humanity. We can understand the drama that moves between positive and negative, but mostly negative. We've played all the parts in this drama, and our planet is moving us on to a better octave of life. We know what negative experiences feel like. This is our contribution to universal consciousness. Negative polarity is life-diminishing and ultimately destructive.

Humanity is now at the destructive point for all negative polarity. At the same time, the Earth is moving beyond the negative, into positive polarity with rising vibratory resonance. This is making all negatively-oriented persons very uncomfortable for no apparent reason. Their consciousness will be guided to a world that has the same energy spectrum as that of most of current humanity.

Chapter 4. Aligning with Higher Consciousness

As we intentionally live in positive vibrations as high as we can imagine and embody, we move into a higher dimension that is unavailable for any negative energetic encounters. Our lives change dramatically. It's as if we just joined a new private club that is amazing in every way, and it's available to everyone whose vibratory level is in alignment. There are already millions of members, and it's easy for us to love one another and be joyful together. These energies become our experiences as we align our thoughts and feelings with them. They're very contagious.

Awakening to Infinite Awareness

We can enter the void of pure awareness only without intrusions from the ego. When we move into pure awareness, the ego cannot help us, because it is limited to the consciousness boundaries of the spectrum of humanity's polarity and vibratory frequencies. It does not know intuitive guidance. Pure awareness it beyond time and space. In our essence we are eternal, unlimited Self-awareness, but the ego cannot fathom this, because it is created to help us navigate within a realm of separation from our eternal Self. In our true Being, we are created to be life-enhancing in our expressions and creations.

We create energetic patterns with the resonance of our thoughts and emotions. We can recognize the quality of all energetic patterns by how we feel about them. Everything that stimulates joy, happiness and fulfillment for all of us, and for every being in our presence, can fill our attention, if we so desire. The ego may be trained to be neutral. Eventually the ego dissolves within our consciousness, because it is no longer needed. We can be in a higher dimension of living by becoming aware of our complete Being, our unlimited, infinitely powerful Creator Self. Each of us can realize this state of Being. It is pure awareness, unlimited in every way and possessing unlimited creative ability. It is also a state of ecstasy and bliss, and it comes to us as much as we are willing to experience. We have control of the focus of our attention and can direct it into the vibratory field of ecstasy and bliss as long as we choose.

Currently we have limiting beliefs that keep us from pure awareness. They're part of our ego consciousness, which we can awaken to the presence of our greater Self-Awareness. It begins with our desire and intent to expand. We learn where we can find positive, elevating environments to help us raise our own vibrations. We can breathe more deeply and rhythmically. We can direct our focus of attention on expanding into present, unlimited awareness. By inviting this level of energy into our awareness, we draw it to us by compatible polarity and resonance. This has required much practice in meditation to achieve. By understanding the energetics involved, we can practice attaining pure awareness, perhaps with more directed attention and greater understanding of the process.

When we are in a state of pure present awareness, the empirical world of humanity becomes recognizable as a limited compartment of human consciousness, which is interpreted in our consciousness as physical experience. We are its creators by aligning with its energetics in our perceptions and feelings, and we can change our experiences at any time by changing our vibratory status. This depends upon our focus of attention and emotional state, both of which we can choose to experience within ourselves, guided by our intuition.

Along with being open to expanding our awareness, we can work on getting clear in our Being, not needing or desiring anything outside of our own consciousness. As we practice, we can become aware of our limitations and resolve them through the higher guidance of our intuition in every moment that we are present in awareness.

Living in Deepest Love and Joy

The life force of the Creator is present for us in every moment. It conveys infinite creative power and unconditional love that enhances our vitality. It provides as much awareness as we are receptive to and capable of realizing. We set our own limits, and we can unlimit ourselves. To do this, we can be completely open to all energies from a perspective of eternal Self-awareness, anticipating greatness of Being. Being completely open to our conscious life force enables us to align with our essential Being in our feelings and thoughts. We can imagine living in a community of fearless, loving and joyful people. Everyone in this community is accomplished at focusing attention on life-enhancing ways of understanding and acting. We can visit and align with the energy of Gaia in beautiful and majestic places.

When we deeply reach for higher awareness, we become aware of out limiting beliefs, and we can resolve them with our intent to pay attention to the energetics that bring us joy and fulfillment. Life can become more intense in every way. The world is brighter, people are kind and happy. Our emotions can go much deeper and richer into the feelings we desire. We can open ourselves to the true energy of the heart of our Being in deepest, passionate ways. It can feel as if we're immersed in ecstasy and bliss as long as we can take it. This is possible by letting go of every feeling of responsibility, stress and fear, and intentionally replacing them with confidence, joy and intuitive guidance, while continuing to live in our chosen situation. This may take much practice, but with strong intention it can be accomplished by any of us.

As much as we can give our sensitivity to our intuition, we can live loving and joyful lives. Intuitive guidance is always present for us, but we must be attentive to it, or we are not aware of it. It is what we truly know. In every moment, this knowing comes to us immediately as we encounter, feel and imagine. We can have a perspective that this is what's happening. In our human experience, we are constantly being guided by our expanded Self. This is the energetic quality that we can align ourselves with. As we recover from living in limited consciousness, we can enjoy life as intensely and joyfully as we can open ourselves to receive with gratitude.

Our Eternal Presence of Awareness

To be aware of our eternal Presence, we can open ourselves to the energy of our higher Self and ask to be drawn into realization of our infinite Being. Once we realize that we are much greater than our human self, we can direct ourselves into positive, high-vibratory visions, feelings and experiences. We can examine our limitations and realize what they are and resolve them. We do not need to be limited.

We can recognize that our human experience is part of a compartmentalized reality, created by the collective recognition of humanity and maintained in our consciousness with our life force. It is designed to reflect back to us the results of our creations. By the focus of our thoughts and emotions, we align with their energetics. Our alignment radiates into the quantum field and manifests the quality of our attention in our experience. This is our creative nature.

Although we are participating in the holographic game of human life, we can realize how to participate in a life-enhancing way that feels wonderful. From the energy that we experience, we learn what we naturally prefer, and we can direct our attention there. We can position ourselves in elevating environments, where we can be naturally thankful and joyful. This level of vibration and polarity is where we can be aware of our intuition with precision. We can know how we feel in great sensitivity with increasing awareness.

In our interactions with others, we can be confident in love and compassion, even in the most difficult encounters. It is always our level of vibration that is important, not the form of what we do. It is by vibratory resonance that we create. We are natural modulators of energetics. We don't have to know any of this in order to be creators. It is our nature. We only need to know that we are creators with our state of being, our thoughts and our emotions, and that we have absolute control of ourselves. We have our imagination, and we can summon up emotions.

We can realize that there are no real threats to us. We're playing roles in a game of consciousness, and we create the quality of our experiences. We have free will in every moment, and in every moment we are creating awareness of experiences for all conscious beings in the universal consciousness of the Creator, which we participate in and can be aware of.

Training for Confronting Challenges

Because of the human condition that we have inherited over many eons, we continue to confront negative energy in our government, our society and ourselves. The lower the vibrations, the more we are stimulated to feel anger, shame, guilt and fear. We try to avoid these energies and feelings, but when we learn to anchor ourselves in positive, high vibrations, it is possible to confront the dark forces and to resolve them with love and compassion, without engaging in alignment with them.

Our experiences depend upon our perspective, which we express through the radiations of our energy signature. If we resist negative energy, we are aligning with it, and we are in the same vibratory spectrum. It does not matter on an energetic level whether we are resisting or supporting it. Resolution can happen only from a higher perspective of positive energy. The polarity must shift, and this can occur only within our own consciousness. Once we make the positive shift, and we maintain it in our perspective, we radiate love and joy and receive experiences that stimulate this level of energy.

To initiate a positive perspective in every moment is our human challenge. It can happen when we release our self-imposed limitations and open ourselves to our eternal presence of awareness. As we resolve our limiting beliefs, we open our awareness to the flood of conscious life force arising from the consciousness of the Creator, which we have been blocking by maintaining our boundaries in human consciousness.

As our awareness opens up, everything becomes more intense, with brightness and new colors and musical sounds and so much more. Beyond the confines of the energetic spectrum of humanity, we naturally have unlimited awareness of all dimensions. We live in the universal consciousness of the Creator and have infinite creative abilities with our thoughts and feelings in alignment with our heart-mind, beyond our ego consciousness.

We create our ego to operate within the energetic spectrum of the empirical world. As we expand our awareness beyond time and space, the ego becomes unnecessary and dissolves for lack of our creative life force. This enhances our personal selfhood, because we become less limited and eventually unlimited. Each of us is the Creator on whatever energetic level we choose to pay attention to. We create experiences for universal consciousness. This is the role of our persona in this life on this planet. We are absolutely free to live however we choose without impinging on the being of another.

As we become more and more positive, with more compassionate and elevating thoughts and emotions, we can recognize the light in everyone, including the parasites. They need to take light from others in order to exist, so the light is present, even though it didn't come through the heart of their being. If we align our energy with our light, we keep our life force and continue to radiate the energy of our heart. We can draw them into resonance with our energy or let them disappear beyond our experience into a different dimension.

Identifying with Our True Essence

We are the only ones who can create our life experiences, and the only ones who can change them, against the backdrop of humanity. We may wish to change others, to make them more acceptable and desirable for us, and the way to do this is to change ourselves. By changing our polarity and vibratory level, we change our perception of others, as well as our perception of ourselves. The big change is moving from fear to love. This results in a major change in perspective and requires strong intention to follow through. Because our ego consciousness has no awareness of our higher guidance, our ego is useless in this process and must be instructed to stand by until called upon.

When we have a positive perspective, we have no doubt or fear that we have unmet needs. As a result, we are given whatever we need for a fulfilled life. Because we have lived entirely by reliance on our ego mind, our ability to calculate and to think inductively and deductively, taking a positive polarity is a major leap in consciousness. Only occasionally have we been aware of our intuition, unless a miracle happens for us through the intervention of our higher Self changing the vibratory patterns in our awareness and allowing us to become aware of a higher dimension of vibrations.

Everything is always available to us, unless we are not open to it. In our culture we have learned to close ourselves off from our higher guidance and to rely on our ego consciousness of limited awareness. We have accepted being in lack and enslaved to negative parasites. When we're ready for it, the path to freedom and fulfillment is within our own Being, by developing sensitivity to our intuitive guidance flowing through the heart of our Being and expressed in our heart-mind. Here is our access to universal consciousness in our own Self, our innermost and deepest Being. It is what we truly know, apart from anything we have been taught or made to believe.

Although our intuitive guidance is always present, we must

develop sensitivity to it by paying attention to our thoughts and feelings when we are unencumbered by ego consciousness. We can ask our guides and angels to stimulate awareness of our intuition. Everything that happens in our lives is for our inner guidance. We are living in a metaphor, a construct of consciousness, that we have created by aligning ourselves with energetic vibratory patterns through our perspective and the way we think and feel about everything.

By being confidently positive and learning to maintain a perspective of compassionate wisdom, we attract experiences that stimulate joy and gratitude in ourselves. We can learn to interact with the inner light in others. By opening our awareness to our inner knowing, and acting confidently from what we deeply know, we eventually gain access to a realm beyond duality and are no longer subject to negative encounters.

Enhancing Our Intuitive Alignment

If we desire to be successful on the inward journey, we must align ourselves with our intuitive knowing to the point that we can be confident in knowing everything we want to know in every moment. This is possible for us. It requires us to be open to being our higher Self, to align intentionally with only positive, high-frequency thoughts and feelings, and to dismiss all distractions from our concerns. To be in the state of being that invites awareness of our intuition, we can learn to discipline ourselves to have complete control over our thoughts and emotions, to be able to call them at will into awareness, and to be able to wield them intentionally, as good actors and actresses.

By controlling our mental and emotional operations, we are able to create and change our vibratory levels at will, allowing us to create whatever we want. When we are aligned with the energy of our heart, our lives are filled with joy and gratitude for the wonderful situations that occur for us. For this to happen, we must remain positive, loving and compassionate in our encounters, real and imaginary. In our true Being, we cannot be threatened or intimidated, because we know intuitively that we are eternal and unlimited creators, and we can master every situation.

We can create ourselves out of any scenario that we desire to leave by refocusing and realigning ourselves with the energy we desire to experience. This may require great intention and confidence in deepest love and gratitude for the light that is within and all around us. Although we need technology to detect this, our environment is filled with light from the presence of photons emitted by all living beings and inherent in the quantum field. Our DNA emits photons, making us beings of light, and we live within our own radiance, which we can recognize, as our awareness expands into our true Being. The brighter we are in high vibrations, the more radiant we become.

Since there is only One consciousness, which we participate in as creators, we can learn to control our state of being by resolving all of our accrued beliefs and fears, so that we can be present in awareness without thought or emotion, just being aware. This can be our normal state of Being, while we go about our lives, guided by our intuition. With our love and compassion, we can contribute to the elevation of humanity's vibratory resonance, while aligning ourselves with the rising vibratory patterns of the Spirit of the Earth, which we can learn to feel, when we're alone in nature. In the wild places, we may become aware of many spirit beings, who express themselves just beyond human perceptive ability, and we may communicate with them in love and joy.

5.

Realizing Our Personal Truth

Living as Our Eternal Selves

Once we are confident in knowing our eternal Being, we are no longer subject to fear of any kind, and we are free to live in the high vibrations of joy and kindness. We are free to imagine every encounter in high vibrations of experience. We can radiate compassion, acceptance, love and assurance of mastery in every situation and encounter with any energy patterns. These are all innate abilities that we all have in our expanded Self. We only need to recognize our higher aspects. It is who we are apart from any other energy patterns, such as human embodiment. It is just being our pure awareness. This is what lives forever. This is where the source of our life is. It is the eternal, Self-Aware universal Creator, the all-encompassing consciousness that constantly creates galaxies and universes and everything.

Although quantum physicists know of this consciousness, they are limited at this point. The nature of consciousness is left to the mystics and spiritual masters to understand and practice. Everything happens by vibrations, which are the expressions of consciousness. Everything in the unified quantum field of all potentialities is an expression of consciousness. We all share in the expression of the empirical world with all of its concepts and feelings. We've been taught to be completely focused within this spectrum of energy. But we can escape just by changing our perspective.

Unified consciousness is unlimited, and it's available to us, if we really want to experience it. It comes through the energy of the Heart and Source of our conscious awareness. It requires complete openness on our part, without any limiting beliefs about ourselves, or any remaining fear of mortality. In expanded conscious awareness, our focus is on high-vibration thoughts and emotions. By living in awareness of this spectrum of energy, whether in our physical experience or in our imagination, we can realize our eternal, unlimited, loving and vivacious Self. We can do this in our imagination, if we practice often. When, because of our practiced ability to be constantly open to high-frequency thoughts and feelings, we finally spontaneously feel and know that we are living in the quality of our visions, they are our real experiences.

Self-Development of Personal Awareness

We are here to develop ourselves to the fullest and to raise the resonant frequencies of humanity into greater light and love. It is what we know deep in our consciousness. It's what we know intuitively in the energy of the heart of our Being. We are at home in the high-frequency vibrations that envelop us. We know the feeling of the deepest, purest love. We know great joy. These are the feelings that guide us into our awareness within universal consciousness. This is where we live in the essence of our true personal Being.

We are our present awareness, expressing ourselves as our personal energy signatures. These manifest as the quality of our personal experiences within the spectrum of the energy frequencies of humanity. This is our current fixation. It is what we are being drawn to transform into alignment with high-frequency thoughts and emotional patterns based in love. We can do this by understanding the flow of life that we experience.

Each of us has predominant thoughts and feelings about our experiences. These thoughts and feelings are constantly creating our qualities of life by their patterns of vibrating frequencies. To raise our vibrations and quality of life, we can use our imagination and emotions to respond to our inner intuitive prompting. We can create what we intuitively feel that we need. We magnetically attract the resonant energy patterns that we recognize as our fulfillment.

We've been here for all the energies we could experience. Our emotions are sensitive to vibrations and always convey to our awareness the nature of the energy confronting us. We know immediately if it is of fear or of love. These are primary distinctions for our recognition. If the energy is of love, we can easily flow with it. If it is of fear, it can be transformed or dissolved. We can transform it by continuing to vibrate at high frequencies, encountering others from a perspective of compassionate wisdom, love, gratitude and joy.

The Hidden Truth of Our Being

Because the empirical world seems so solid and real to us, we have been largely unaware of its true nature and how our experiences transpire. Only after the discovery of quantum physics did scientists realize that everything is actually relative to our perceptions. Our perceptions create our experiences and our reality.

Our world and the entire cosmos that we inhabit consist of a vast plasma field of electromagnetic energy and patterns of energy emitted from universal consciousness, in which we participate. As a population of humans, we all agree to recognize the spectrum of energy known to us as the empirical world. Within this spectrum of vibrations, all of our thoughts, emotions and experiences are expressed. In our conscious awareness we recognize all of these vibrations, but we have been largely unaware of the vast sea of energy beyond the empirical. Our enclosed spectrum of energetic patterns has served us in allowing us to participate in experiences that would be impossible in our larger consciousness. In our true nature we could not know suffering and fear. We could not have the depth of experience that now gives us deep compassionate understanding and greater love for one another and our Creator essence.

We can now be thankful for the eons of low-vibration experience we've endured in shaping our character, as well as those who have played the part of the dark masters and their minions. We have been challenged to realize our true Being, and never more urgently than now. As the resonant frequency of our galactic environment and of our planet rise in response to the blasts of high-frequency gamma ray photons that we are receiving, our physical bodies and our energetic presence are also being stimulated to rise in frequency and become more loving and compassionate, as well as more intentionally creative.

We are at the turning of the eras, and a new world of beauty and joy awaits us, as we begin to participate in its creation in our own experience. We accomplish this in our consciousness, in our imagination and emotions. We become expressions of the energy we wish to experience. As we become predominantly compassionate and loving in our interactions, our world of experience changes to align with our rising vibrations.

Deepest Love in Infinite Being

In the spectrum of human energy that we have chosen to inhabit, we have developed beliefs that keep us from wandering outside of our accepted energetic patterns. We've learned to focus on frequencies that stimulate discomfort and fear, even though this is not what we want.

We may feel that we want more abundance, but this can result in containing ourselves in low-frequency energy. Even the most successful among us often indulge in financial investments, such as the stock, bond and commodities markets, which are zero-sum games, in which someone who gains does so at the expense of someone who loses. We try to outwit each other in competitive strategies. We cannot all get ahead in zero-sum investments. Even investing in things that increase in monetary value due to inflation is theft from everyone else who depends upon the constant value of the currency. It appears that we cannot be successful without stepping on someone on the other side of our position. Social status is another way of gaining success at the expense of the less successful, and they are given life force by those who hold them in a higher status. This inherently feels like low-vibration energy. There is another way, and it depends on miraculous happenings.

Celebrities receive a tremendous amount of life force in front of thousands of adoring fans. This is the scale of life force that we all have available ourselves, once we are completely in the flow with the energy spectrum at the heart of our Being. This is where our awareness begins in complete clarity. The more we search and strive for inner clarity, the more balanced and objective we become. The objective here is to continue to pay attention to vibrations that feel like love and abundance, regardless of anything we may be experiencing among humanity. When we can maintain this focus, without any attraction to low-vibration intrusions, we become observers, as our life decisions are guided by the intuition from the source of our awareness.

The I behind our awareness is our true essence, the source of our conscious life. In the vibrations of our true Self, we experience the most wonderful ecstasy of Being in eternal, unconditional love and joy. These are our natural vibrations and are very comfortable for us. Once we have left fear-based vibrations behind and are no longer influenced by their frequency patterns, we can naturally live in abundance in every way. We believe and know that we are constantly creating the qualities of our reality by where we direct our life force through our attention and emotional alignment. In our clarity, we can know in our intuition everything we want to know and imagine. The energy patterns that we choose to focus upon also receive our emotional alignment. This alignment creates the vibration of our personal energy signature. It is the expression of our true Self filtered through out beliefs. Without beliefs in our limitations, we can realize the unlimited conscious Self-Awareness of our Creator.

Opening Ourselves to Higher Guidance

In our true essence we are self-conscious Beings, arising out of the universal consciousness of the Creator of all. We are fractals of the One, having all attributes and abilities of the Creator. In our essence each of us is the Creator.

Prior to our incarnation on this planet, we desired to know even deeper love and compassion than we had ever known, and we decided to gain this depth of knowing by experiencing its opposite. We compartmentalized our consciousness into the limitations of the empirical spectrum of energetic vibrations so that we could lower our own vibrations into the level of fear and suffering, believing ourselves to be mortal. We created our ego consciousness to navigate this realm, guided by our limited mental processes and social conditioning. And we intended to awaken out of our compartmentalized hypnotic trance to our intuitive knowing in order to return to our true Being in infinite love and joy.

In the human experience we have chosen to know the entire spectrum of low vibrations, all of which stimulate some level of anger, judgment, fear and stress in us. In these vibrations love does not exist in its true essence. It becomes conditional attachment. Our true essence is unconditional love, and when we do not realize its presence in us, we feel deprived and needy. We imagine all manner of deprivations, such as poverty, hunger and fear of survival. These conditions are all self-created, many of them sub-consciously, as a result of our predominant perspective of low vibrations. When we are in the midst of low-frequency experiences, we become emotionally unable to choose our desired higher-vibrational state of being. We keep feeling and recognizing our distraught condition and creating it over and over by our continuing focus, sending it our life force.

Only when we realize that we are capable of true unconditional love can we begin to extract ourselves from vibrating at low frequencies. We can recognize that everything we experience has a purpose, ultimately to allow us to acquire the deepest compassionate wisdom. We can gain access to high-frequency love vibrations in the energy of the heart of our Being. If we make this our journey into the truth of ourselves, we can transform our lives into continuing experiences of love and abundance. We do this by raising our vibrations and recognizing the light in the heart of everyone, even the darkest ones. It is this light that we can relate to, and only to this light from the infinite One. We can open our awareness to a higher dimension of life that is always present for us, just awaiting our recognition and the realization of our presence in it.

Aligning with Expanding Consciousness

In order to be able to live more beautiful and abundant lives, we must search within our own consciousness for the source of our life force and intuitive knowing. Our feelings can guide us there. We feel the vibrations of everything we encounter. We know when we're in the presence of love or fear. These are the polarities that we move between in our psyche. Each of them has a spectrum of frequencies that feel like the vibrations that we recognize them to be. We make decisions all the time to vibrate at our chosen frequency of love or fear. This level of our conscious vibratory expression determines the quality of our experiences.

By choosing to focus within the spectrum of love, compassion, freedom, gratitude, joy and abundance, we flow toward greater realization of our unlimited, fully-conscious Self. In this spectrum of vibrations, we can realize our eternal presence of Being in the universal consciousness of the Creator of all. In this realization, we know the truth of who we are. Once we experience this state of Being, it remains with us, because it is indelible. It is our natural state of Being, with all of our natural abilities of creation.

In our expanded Self of present awareness, we cannot be threatened successfully, because we have learned how to transform any energy patterns we encounter. We can embody the energy of our eternal Selves and manifest our physical presence in mastery of all encounters and situations.

We cannot be in our higher Selves while we are focused and aligned with the dramas and threats of human experience. We can observe them and interact with them, while holding and expressing compassionate wisdom. In this perspective, we can radiate our alignment in higher-vibratory energy patterns. The low vibrations of all of the destructive energy patterns in our awareness will experience interference in their frequencies, become unstable and disappear into the vacuum fluctuation of the quantum field.

Our interaction with energy patterns within our consciousness becomes easier and more proficient as the resonant frequency of the Spirit of the Earth rises, along with all the Beings in our galaxy and beyond. As we vibrate higher, this process can be immediate, but now it may still need some time.

Low-vibration, negative polarity energy has been predominant here, but the electromagnetic polarity has shifted, and now we are flowing in a positive polarity with rising vibratory frequency. We can realize a higher dimension of living in greater love and fulfillment, while attracting all who are intentionally searching for inner truth to come into alignment with higher vibrations.

Flowing into Higher Consciousness

Our compartmentalized human consciousness has kept us from realizing our greater Selves. We could not allow ourselves to know what we don't believe is possible. The resonant frequency of the Earth is rising in waves of higher vibrations, summoning human consciousness to rise into greater realization of our inner light. From this source of our conscious Being, we constantly release photons that create our radiant aura of etheric colors that some of us can see. In the near future, all of us will come into alignment with these higher vibrations in our feelings and thoughts, or we will be very uncomfortable and unstable. It will be difficult for anyone who continues to resonate with low-frequency thoughts and emotions.

When we allow ourselves to flow with the higher vibrations that are streaming into us now, we find that our vitality is rising, and we are being guided into life-enhancing energy patterns. We can experience deeper love and awareness of our eternal consciousness. As we continue to penetrate and expand our awareness, we can begin to enter into universal consciousness, the Being of our Creator. This is the highest level of vibrations and is like facing close to the sun and having every vibration lower than divine, unconditional love and joy burned away, leaving us clear and vibrant, able to create intentionally anything that is life-enhancing for all involved.

We can also ease into higher consciousness more gently, but it still requires strong intention to proceed on the journey into the heart of our Being. All of our limiting beliefs about ourselves will confront us. We will need to resolve them through acceptance, understanding, compassion and transformation or dissolution. As much as possible we can imagine and feel ourselves being our higher-vibratory Self. We know what feelings are stimulated by all of the vibrations, and we can entertain scenarios of great love and joy. These raise our energy signature to higher vibrations, attracting more love into our lives and transforming our experiences. We can enter into higher vibrations than the best romance novels could ever portray. It all happens within our own awareness, as we open ourselves to higher and higher inspired vibrations, feeding us joy and ecstasy. We can live in this higher spectrum of energy, while also being grounded with the Earth, where we can find elevating energies transforming our experiences.

Living in An Ascended State

We have the ability to radiate so much light, that our presence can be blinding to anyone aligned with low vibrations of fear and lust. They either transform into alignment with love, or they dissolve into a lower-frequency dimension. We can be the Presence of the Divine One, I AM, fully open to the highest frequencies of positive polarity vibrations of awareness in the universal consciousness of the Creator.

In our human spectrum of operating frequencies, we can hardly imagine that state of Being, but somehow deep within, we know it faintly. It attracts our attention, because it feels so wonderful and vibrant. We may have to search for it, so that we can realize our potential. It requires our going deeper and deeper into love and joy, higher and higher vibrations with more intensity in amplitude.

As the conscious creators of our experiences, we can learn to use our imagination and emotions to align with higher vibratory experiences with more freedom and sovereignty. As we do this, our energetic signature rises in frequency patterns, attracting higher vibrational energy patterns that align with us. Because we are energy modulators, we can change the frequency patterns of any energy we encounter with our thoughts and feelings. When we are in a state of clarity in our intentions, we can expand our awareness beyond our limitations of space and time. We can realize the eternal awareness of our presence of Being.

We are fractals of the Creator, having all of the awareness of the Creator beyond our personhood. Our abilities are unlimited in every way. Our awareness can be unlimited in every present moment. The higher in joy that we can go, the deeper we are going into universal consciousness. As this experience grows in intensity, our energetic frequency increases, moving us into a higher dimension of living. We can use our imagination to open ourselves to this. Our elevating feelings can keep us in the vibe of deepest love and compassionate wisdom.

From the perspective of higher guidance through our intuition, we are prompted to master our circumstances in whatever situation we recognize ourselves experiencing, whether physically or imaginatively. When threatened, instead of reacting in fear, we can react in love, expecting to interact in a kind and loving way. We do not need to engage and align with low-vibratory energy patterns. Through our inner guidance we can transform or dissolve them, when we are in a state of objective clarity beyond the limitations of compartmentalized human consciousness.

Being the One Conscious Creator

In our true Being we are many persons in many dimensions simultaneously beyond time and space. Quantum physics has demonstrated that subatomic wave patterns can be in many places concurrently, until we observe them, when they become part of our time-space continuum and materialize for us. By logical extension, this applies to our presence here as well. Our personal beings are limited only by our own self-imposed constraints. Most of our other-dimensional persons are not consciously compartmentalized the way we are as Earth humans.

Having subjected ourselves to the spectrum of human consciousness, we have voluntarily closed off the awareness of our higher Self. We have largely lost touch with our intuitive knowing. We have allowed ourselves to believe that we are sinful and deserve to be enslaved. Many have allowed themselves even to be injected with nanobots in preparation for transhumanism, from which there can be no escape. We are at the turning point in history, when each of us must decide either to continue to allow ourselves to be enslaved even more drastically or to awaken to our true Being. There is no other choice. If we want to transform our lives, we must learn to use our consciousness as intended by our Creator.

Once we decide to transform ourselves, we enter the inward path to the energy of the heart of our Being. No one can take us there. It is a solo, personal journey of awakening. It cannot fully be taught; it can only be known. We can become acquainted with our true Innate Being, who serves us in compassionate wisdom and unconditional love and brings us every experience that we believe in and need for our personal growth and development as Beings of light and love.

We are being prompted by the current in the quantum field, which envelops us, to raise our vibratory frequency from the negative polarity, low frequency perspective of helplessness to the greatest expansion of our conscious awareness that we can imagine and feel. This process involves resolving all of our limiting beliefs of inferiority, lack and ultimate mortality in order to realize the eternal, unlimited creative power of our true Self.

When we begin to realize our truth, we can allow ourselves to trust our higher guidance. We can have absolute control of our life situations, regardless of everything that appears to be happening outside of us. All of that is actually a reflection of our own conscious processes. By intentionally living in gratitude and compassion, we elevate our vibrations and magically transform our lives.

Expanding Our Conscious Awareness

Expanded awareness in higher dimensions is our natural state of Being. In our true Being we are multi-dimensional. Everything that keeps us from realizing our eternal personal presence of Being, is self-imposed. In order to experience the spectrum of energy that is known to humans, we had to compartmentalize our consciousness by intentionally or subconsciously instilling in ourselves the belief that we are mortal and subject to all kinds of low-vibration environments that want our life force through our alignment with their resonant vibrations. If we recognize them from a higher dimensional perspective, they cannot be our parasites. They need our permission to take our life force. We have succumbed to their low-vibration enticements, but we are not required to stay at our current level of vibrations.

We are designed to be the masters of our energetic expressions that become the qualities of our experiences. This creative ability that we have has been known only by esoteric groups, but the vibratory level of Gaia and the growing number of awakening persons have attained a level of consciousness in which this knowledge can be released to the public.

While humanity was stuck in fear and ignorant of our all-knowing inner guidance, we had to rely on our ego-consciousness to navigate the human dimension without higher guidance. We have been playing roles in a play that we designed, directed and produced without knowing this lesson. The lesson is to recognize that we are the creators. Once we realize this, we can know the importance of lovingly directing our focus on the level of energetic patterns that we want to live in.

In interacting with others of every kind, we can face everyone from a perspective of compassionate wisdom, remaining in a state of high-frequency thoughts and feelings. We can realize that we are eternal, Self-realized Beings, and we've been role-playing in the human drama. Our human body is a complex pattern of electromagnetic waves that we perceive and believe is our physical body. We hold its creation and state of being in our own consciousness, and we can elevate our state of being with our intentional focus of attention on feelings and thoughts that are wonderful for us in expanding our capacity to be loving and compassionate.

Once we know that we are eternal, personal and present conscious awareness, we can be objective about our human life and learn the real lessons. The most significant lesson is to learn how to change our polarity from a negative, low-vibratory level of fear, to a positive, high-vibratory level of love. These are energetic opposites, and they can interact, but not align. Their frequency and polarity are mis-aligned. There is discomfort and disharmony, resulting in separation.

It is a leap in consciousness to move from fear to love. We can do it by opening ourselves to the truth of our own Being. We can become sensitive to our intuition by searching ourselves for the source of our conscious life force and being present with it. We can intend to feel the unconditional love within the life-stream expressed by the heart of our Being. This is beyond the physical body. When we're ready, we can recognize our unlimited Selves in our conscious awareness. Once we know who we are, our interactions among other humans become heart-felt and thankful. We can appreciate the intricate web of energetic patterns that we have all created, and we can elevate them in our own experience.

Awareness of Self-Realization

As a race, we have become disconnected from awareness and

care of all of the other conscious beings around us, including our planet. The first step toward enlightenment is to reestablish that connection and begin to recognize that we are all the same Being, just different aspects of the One. When we feel that we have developed deep compassion and have realized our natural attraction to all other conscious beings, we have completed our lessons in the dimension of low-frequency, human vibrations. We can graduate to the next level, which is a higher dimension with a different polarity and higher frequency of vibrations. We need to shift from negative to positive polarity, from life-diminishing to life-enhancing vibrations. This creates a different perspective with much greater awareness of the expansiveness of life and the choices to be recognized.

If we decide to take the leap of changing polarities and frequencies, our life changes dramatically, probably over some time, because we may need practice to be successful. The key element is resonant vibration. We can learn to stay in intentional alignment with the Source of our conscious life force, which is the source of our creative abilities.

We can encounter the heart of our Being, which is our inner light, the source of our photonic presence, which radiates through every cell of our body and forms our aura of colorful, conscious photons. We can make this encounter intentionally by seeking higher emotional and mental vibrations and by just being present in awareness. This is the place of realization of our eternal present awareness beyond time and space. We have always been and always will be Self-aware.

As we raise the frequency of our energy signature through focusing our attention on high-vibration thoughts and feelings, we begin to feel the presence of Creator consciousness. We can open ourselves to higher frequencies of greater gratitude and compassion, as we reorient ourselves to deeper and more meaningful love and sovereignty.

The Attraction of the Inward Journey

We designed the human experience to disable our connection to higher consciousness in order to make our lives here dramatic and impactful. We wanted to gain deep compassion and understanding of a greater spectrum of awareness than we could have as higher-dimensional Beings. In order to become aware of our true identity, we must intentionally achieve it from within the human compartmentalized consciousness.

Our abilities for this inner journey are our creative imagination and our emotions. We are also the creators of the quality of our experiences through our modulation of energy patterns with our thoughts and emotions. At some point in our personal evolution, we feel that there is more to life than we have been aware of. We begin to become aware of greater compassion and understanding. We begin to notice how we feel about everything, and we become attracted to more life-enhancing situations and people.

As we open ourselves to higher-frequency, positive experiences, our lives begin to feel better. There's more love and kindness in everyone around us and more abundance and freedom in our lives. It all depends upon how we filter our awareness with our beliefs and perspectives. There is always a higher truth to recognize in every situation. Resonant frequencies occur in octaves, and there are vibrational patterns in a higher dimensional spectrum in every aspect of life. We can be aware of these levels.

Realizations come from the consciousness at the heart of our Being, when we are open to them. This is our intuitive knowing, and it is our higher Self guiding us in the way that we can best understand. Because we have lived within a compartment of our consciousness, we have to recognize our self-imposed limitations and resolve them.

We can be unlimited in our imagination and feelings. By intentionally imagining and feeling ourselves living in scenarios that feel really good with high vibrations, we become able to transform the polarity in each situation from negative to positive and maintain a perspective of compassionate wisdom and joy. By elevating our own energy signature, we also elevate the frequency levels of humanity, as our personal radiance grows in vitality and clarity. As our sense of compassion grows, we gradually move into awareness of the singularity of universal consciousness. We are all One Being, acting out our roles for the greatest experiences of all.

Our Path to Awakening

Part of our participation in the human experience is our training and programming from infancy on to the present. We've been taught that we are limited in our abilities, that we live in a world of limited resources, and that we must compete with one another for success. We've also been taught that we are inherently aberrant and in need of control by government and absolution by the church. If we analyze each of these perspectives, we find that none of them can possibly be true.

From Quantum physics, we know that there is a universal consciousness that creates and bestows consciousness to everyone and everything that exists. The entire cosmos works harmoniously and in perfect order, guided by great Beings of light and love. All creative impulses are life-enhancing, engendering love, joy and freedom. Freedom requires free will, and it is the free will of some beings that has resulted in our anomaly. We decided to experience something that would be impossible for us in our naturally-created state of unlimited consciousness.

We created a limited compartment of our consciousness in order to experiment with negative polarity, low-frequency thoughts and emotions. We entered the realm of darkness and evil, in order to know what these conditions are like. Because of our training and the circumstances that we have been forced to live under, we appear to be imperfect and anomalous. This is due to our ignorance in the use of our free will. We knew that our experience would be difficult and life-diminishing, and we designed it so that at the appropriate time, we would be able to open our awareness to our true Being, and to have much deeper compassion and understanding than would have been possible without this experience.

That time is now. Humanity is awakening to our greater consciousness. We are on the verge of becoming galactic Beings of love and light. Our consciousness is universal. Our expanded awareness is the consciousness of the Creator, whom we can know through the conscious life force flowing through the heart of our Being, and focused in our intuition.

By intentionally breathing deeply and rhythmically and asking for the feeling and awareness of this energy, we become conscious of it. We begin to free ourselves from the outer situations that we have believed we are in, and we gain the ability to create our experiences from within.

Consequences of Free-Will Choice

We create, direct and act out our lives according to the qualities of our intentional and unintentional thoughts and feelings in each moment. These are the expressions that arise out of our predominant emotional and mental polarity and vibratory levels. The essence of our creative ability is the way that we modulate electromagnetic wave patterns.

Every experience is an encounter with electromagnetic waves and patterns of waves. We are endowed with free will and have the choice of which wave patterns we recognize. These become the reality we experience. Our personal energetic signature is radiant and projects its vibratory resonance into the quantum field, out of which everything arises and disappears trillions of times each second. As our own energetics change, so do our creations and experiences. The situations around us naturally change in quality as our energetics express them.

We create what we longingly or fearfully dwell upon very much. It is what we believe is happening and what will happen. There is no difference in the creative power of our fears or our desires. It all gets filtered through our beliefs, submissions and intentions. By intentionally choosing a level of vibration to focus upon and feel, we can train our subconscious to resonate in alignment. Through repetition or sudden shock, we can create energetic pathways that become predominant in our consciousness. These are the changing components of our personality.

We are playing roles within a humanly created realm that we have been programmed to believe is empirical and real, and that we do not exist outside of the boundaries of this realm. We're here to learn about negative polarity experiences in order to expand our consciousness into a realm that we could not imagine otherwise. When we've had enough of this dimension, we begin to look for something more expanded, and we start to feel our intuition more strongly.

As we become sensitive to our inner knowing, we recognize that we prefer a positive polarity. We would rather live in love than in fear. We begin to transform our lives. We no longer need to suffer or fear death. We can begin to recognize that, in our essence, we are pure Self-aware consciousness. We exist in the eternal ever-present moment. We can express ourselves in any body or dimension that we can feel in alignment with. We are completely free and sovereign, able to create any experience that we choose.

Knowing Our Invulnerability

Having been programmed to believe that we are our material body, we have believed that we gradually lose vitality and then terminate. We have believed that we are mortal. This has enabled us to have access to negative, low-vibration experiences that we believe are real, in order to deepen our understanding and compassion.

As we become aware of being beyond our body-consciousness, as we do in day-dreams and night-dreams, we can deduce that we are more than our body. There are hundreds of easy-to-find accounts from people who have died and returned to tell of their expanded awareness in being without a body.

In fact, we are so immortal, that even if we strongly believe that we are mortal, this belief has no effect, because we are extensions of universal consciousness beyond time and space. Even in our dreams, we can be many different forms and personal expressions. If we are eternal in our Being, that doesn't mean that our body is eternal, because it is an expression of our beliefs, if we still have them, and our recognition through the focus of our attention. Our conscious self-image and self-beliefs are modulators of the energy patterns that attract our experiences.

Our activities, involvements and all empirical experiences are a result of the vibratory level of what we think about and feel. Everything else results from our resonant frequency. Forms and scenarios arise out of the quantum field, attracted by energy patterns that they align with. These become our experiences.

We are the conscious creators of the quality of our experiences by the polarity and vibratory frequency of our energy signature, which can change moment-to-moment, but has a predominant vibration, in alignment with our perspective on life.

Once we know and believe that we are eternal Beings, we understand and feel that we are sovereign, fearless and free. Outer circumstances rearrange themselves to our changing perspective and quality of life. When we are thoroughly positive and of high vibration, we can expand our awareness beyond any limitations. It becomes possible to live in the body as well as beyond at the same time. The dimensions interpenetrate at different polarities and frequencies.

The Power of Radiant Love

We live in a vast sea of vibrations of electromagnetic wave patterns. The spectrum of recognition in our human embodiment is minute, compared to the vastness of the quantum field of all potentialities, yet we live in our conscious awareness within this frequency range that we hold in our perceptions, as if we are trapped within a prescribed sphere of consciousness. We are experiencing this realm by our own consent, since in our true Being, we are sovereign and free in every way.

In our true essence, we are unlimited and can realize the totality of cosmic consciousness. In order to do this, we must penetrate the boundaries and limitations that we have imposed upon ourselves in order to live in the human dimension. In this era of transition into expanded consciousness, all of our personal dramas and concerns about our relationships and life-situations are distractions from realizing who we truly are. We have been given the gift of potential personal transformation through lockdowns and deprivation to transcend our accustomed striving to survive and thrive as humans in a negatively-polarized environment.

By searching within our own Being, we can know that the consciousness that we participate in is unlimited. Our imagination can encompass the entire cosmos, and our emotions can express the depths of despair to the heights of ecstasy. We may not even know how far we can go with our realizations. Each of us is a potential human prodigy and genius in our imagination and emotions. We each can have a unique expression of our creative abilities in positive, life-enhancing ways in alignment with the intuitive knowing that comes to us through the heart of our Being.

We are constantly receiving a stream of conscious life force that is unlimited, flowing out of the universal consciousness of the Creator of all. By opening ourselves to positive, life-enhancing visions and feelings, we can enter the consciousness of the Creator. We can realize our eternal presence of awareness flowing in unconditional love in unity with all conscious beings, including those intent on destruction. They are anomalous expressions of us, and they also receive divine life force. With focused intent, beyond our self-imposed personal beliefs and limitations, we can realize the unlimited consciousness that is our natural state of Being. With this realization, while maintaining a positive, high-frequency state of being, we can master any situation within the human spectrum of limited consciousness.

Our Potential Awareness

As fractals of the consciousness of the Creator, the potential of each of us is beyond superhero-status. Our conscious vitality constantly inspires us to adjust our vibrations higher. This is universal consciousness drawing us into alignment with Creator energy. As humans, we have blocked our awareness of this. Our ego-consciousness cannot know higher guidance. We have taken on this perspective in order to participate in negative, low-vibration, dark experiences and enslavement. From the perspective of our true Being, we could not imagine what dark energy feels like, so we decided to deepen our understanding. In order to make the human experience real for us, we had to compartmentalize our consciousness with limiting beliefs about ourselves.

We can become aware of our limitations, if we want to. Once we recognize them for what they are, we can resolve them by transforming their energy through forgiveness, love and recognition of their unnatural existence. Their energy is not of the Creator's essence. They exist only by human thought-forms and fixations. We are not required to participate in this band of energy. As we realize the nature and purpose of human life, we can open our awareness to a greater sense of Being. We can learn to be sensitive to our intuitive knowing in every moment.

Intuition is the greatest asset that we have. It can align us with a higher dimension of living by drawing our awareness into greater love, joy and abundance in all aspects of life. It is aligned with the conscious life force that enlightens and enlivens us, and it pervades our Being.

By living in constant gratitude and expectation of visionary experiences, we can align with the most life-enhancing flow of life force through our choice of awareness. We can express our free will at all times through the focus of our attention and emotional alignment. By paying attention to high-vibratory visions and feelings, we can transform our lives into a higher, more refined dimension. We can practice this until it becomes our reality, which it does as soon as we realize it through our inner knowing.

In our conscious awareness we can go between the time-limited and the timeless realms at will. Our awareness of our timeless essence of Being allows us to resolve all limitations in our time-limited consciousness. We can open ourselves to the full realization of our true Being in the unlimited conscious awareness of the Creator.

Repolarizing the World Outside and Inside

The world that we believe is outside of ourselves in our human experience, is actually within our own consciousness. It manifests the quality of vibration in our own awareness of who we believe we are in every moment. The polarity of our focus and alignment is the vibratory pattern that we experience in our human lives.

The world that we believe is outside of ourselves could be grim and threatening, and we are not required to participate in that level of energy. Our vibratory level determines the quality of our personal experience in any kind of situation. Everything is formed in consciousness, which we participate in. We have the ability to direct our focus to positive, high-vibratory feelings and visions. Our intuition presents these to us, if we seek them. If we can hold this focus, we can confront every seemingly challenging outer experience with creative modulation of positive polarity. We can receive inspiration in every moment from our intuition, which offers us positive guidance immediately as needed. It is the first thought and feeling that we receive in any confrontation. If we can stay centered and balanced in objective clarity, we always know how to be in a positive, loving and joyful state, able to act compassionately and wisely with personal power.

Since every scenario is within our own consciousness, it can be positive with high vibrations, even if it appears threatening. With the new Earth energies, the negative no longer has any effect on us, except what we give it. By aligning with the vibratory resonance of Gaia, we participate in the dominant energetic patterns of our planet. These are positive electrical and magnetic wave patterns that affect our thoughts and emotions. By being positive, we are supported by the resonance of Gaia. The negative polarity on Earth is weakening as humanity withdraws its alignment from it, and it receives no life force from Gaia.

By focusing on knowing our intuition, we can make great strides on the inner path. Once we can easily and clearly feel and know our intuitive guidance and have absolute confidence in it, we can live successfully without attachment in every moment. This is a time for withdrawing some of our attention from the persona we've been playing and opening our awareness to our expansive Selves.

Without judgment, we can have compassion and love for all, including those playing the part of evildoers. It is all for our experience and education in aligning with our true eternal Selves.

Death and Transformation

In contemplating the meaning of death, it may be helpful to contemplate the meaning of birth. We are conscious beings, aware of our own nature and able to change everything about ourselves by the intentional focus of our attention. Incarnation as a human being requires this focus within the empirical spectrum of vibrations in order to continue to realize its reality. Prior to our human self-awareness, our physical life begins as a collection of experiences and innate consciousness responses. Once we become aware of our personal awareness, we can enter the school of life. Here we learn what all of the vibrations associated with the empirical band of frequencies feel like when we recognize them as real. After our experience with many patterns of energy, both positive and negative, we develop a perspective of our preferences and beliefs about our reality. Our limiting beliefs about ourselves have kept us from knowing our unlimited nature.

We can open our awareness to whatever we want to imagine and feel. If we are positive and love-centered in our intentions, we raise our vibrations and have better life experiences. Just being open to deepest love, gratitude and joy brings us into a state of being that elevates our personal lives and expands our awareness. We can learn to realize ourselves as high-vibratory positive Beings of great inner light.

Apart from the empirical realm, we can learn to be our present awareness. We can be as unlimited as we are willing to imagine and feel. In our consciousness, we can go as casually or as deeply as we want with any energy patterns in every moment. When a limiting belief comes up, we can recognize it for what it is and resolve it with compassionate wisdom.

Our subconscious or innate consciousness controls all aspects of our body and is the repository of all personal karma. Its intention is to provide experiences that will challenge us to become clear about ourselves. We can learn to be present in our awareness and to allow whatever arises with acceptance, gratitude, compassion and wisdom. This is a practice that requires intentional awareness of our ever-present intuition. Here we receive the high-frequency divine guidance that arises from universal consciousness. It comes through our innate being, which is a non-judgmental consciousness. We can learn to interpret the prompts that we receive intuitively. It is mostly a matter of staying in a state of high-frequency, positive vibrations.

If we have learned all of this, we may graduate to our natural state of unlimited awareness in alignment with the consciousness of the Creator. At this point we are not subject to death in any form, including that of the body. With the help of our innate being, we can transform the physical into a higher dimensional entity with an improved DNA. This opportunity is now present for us.

Awakening Our Soul Potential

Cosmic energies are prompting us to realize a higher dimension of living, awakening in us greater compassion and awareness of our divine Being. The energetic spectrum that we have been living in is being transformed from a fear-based realm to a world of love and abundance. Our living planet is transforming into a sphere of beauty, peace and harmony. Being drawn by the consciousness of great Beings of light, the energy field of human consciousness is expanding into greater awareness of our true nature as fractals of the consciousness of the supreme Creator.

This is a time of coming together in unity of spirit and celebration of our eternal Being in great joy and deepest love. Harmonized by our beliefs, we create the qualities of our lives through the vibrations that we resonate at. As our worst fears and anxieties are being drawn into our awareness, we can recognize them in the light of our true Being. We have been deeply scarred and hurt, because we did not know our real nature, but now we can awaken to abilities that have waited in the depths of our being for this time of enlightenment.

Our experiment of living in a compartmentalized consciousness of negative polarity and life-diminishing perspectives is coming to an end. Through our intentional expansion of awareness, we can reach out emotionally and mentally into the energetic level of joy and abundance. These vibrations have already been anchored and established by higher beings. All we have to do is identify with them and allow our inner transformation to develop.

No longer do we need our self-imposed and accepted limiting beliefs about how inadequate we have believed ourselves to be. We are constantly being created as expressions of the entire Being of the Creator of all, and we have all of the creative abilities of our Source, Who experiences everything that we experience.

Regardless of how much pressure may be apparently imposed upon us from outside, our vibratory frequency is always our personal choice. What seems to be outside of us is actually within our own consciousness. We have lived in a kind of hypnotic trance, dreaming our material world experience. As we awaken to our greater reality in unconditional love and universal consciousness, we can assume our position as the Creators that we have always been meant to be.

Encouraging Greater Awareness

We can live in the greatest love and highest joy that we can envision and feel. We can be in any emotional and mental state that we focus our attention on. Thoughts and emotions vibrate in synchronous electromagnetic wave patterns, with thoughts being electrically charged and emotions being magnetically polarized in the same frequencies. We think and feel at the same time. These are expressions of our awareness.

We have learned to react to situations that appear to be outside of our own being. In actuality there is no outside of our awareness, only outside of our body. If we want to enjoy greater awareness, we must drop our attachments to everything and everyone. We must trust in our eternal Being and infinite abilities. We are all the same Being, sharing the same conscious life force and the same universal consciousness. We are created to be creators. We have the ability to change the wave patterns and frequencies in our awareness by our perspective.

We can understand any situation in any way that we want, and it becomes true for us. If we are in fear, anxiety or torment, we are negatively polarized, and we experience this perspective. If we are in a state of love and joy, we are positively polarized, and we vibrate with the enhancement of life within our awareness. These states of being can be intentional for us, and we can intend to stay positive, even in challenging situations. If we engage and align with negativity, it may stimulate great fear. In the present we are always creating the quality of our life experiences by our thoughts and emotions, and by the vibratory level of our words and actions.

By our polarity and vibratory frequency, we create the quality of our experiences. Since we are all within the One consciousness, the circumstances of our experiences arrange themselves to accommodate our state of being, which we express as our energy signature, in every moment. We can even move beyond karma in our awareness and resulting life experiences. To do this we must resolve all limiting beliefs, including those hidden deeply within our subconscious innate being. When we have a strong intention to know the truth about our Being, our intuition guides us through the process of opening to greater awareness. When we need guidance, we can learn to be aware of the first feeling/thought that we receive in every moment. We can focus our attention on the best, most life-enhancing feelings and scenarios that we can imagine in any situation.

Recovering Our Creative Essence

Our imagination and our emotions are the most powerful assets of our conscious Being. We're accustomed to using them reactively and carelessly, not knowing their true purpose and capability. Our Being is constantly created in love and expectation of greater experience in all dimensions of living. In our essential Self, we arise from the consciousness that creates everything. We are the offspring of the Creator, with full participation in universal consciousness, whenever we desire and allow it in our current personhood. Our imagination and emotions have given us the quality of every experience we've had. Many of our subconscious fixations are designed to keep us focused within the empirical human perspective of reality.

If we can loosen our grasp on all of the energy patterns that are diminishing of our life, we can expand into greater awareness with more compassion and gratitude for deeper understanding. We are the creative essence in our personhood. Our purpose is to use the freedom of our choice of focused attention for creating experiences that we are interested in. By how we live, we are expanding experience for universal consciousness. The experiences that we have as humans in our current dimension are powerful in their energetic qualities.

We've participated in a realm without much wisdom or true love. At any time we can decide to try something better. This is when our true abilities can come into our awareness. We can decide to be positive always, regardless of outer circumstances. This is the vibratory level of joy and gratitude. In this perspective we can receive every encounter and understand what it's about. Then we can bring it into alignment with our vibratory state of being or withdraw our focus and let it disappear into another dimension, as we focus upon the energetic pattern that we want to align with.

We live in many dimensions beyond space and time in the consciousness of the Creator. The most wonderful energy in our human presence is symbolized by our heart, which lives to enliven us and guide our vibratory level in positive polarity. It is the closest comparison to the unconditional love that streams to us in our conscious life force and is beyond limited conscious awareness. Once we unlimit ourselves, we can realize it in our unity with all beings everywhere. We share the same conscious life force and are all conceived and connected in love and joy. These are the emotions of our eternal, unlimited Self-Awareness. In this vibratory level, we can be powerful, consciously intent creators in every moment.

Balancing Negative and Positive Energetics of Life

As humans we have agreed to participate in life experiences within a limited spectrum of energies in order to know what it is like to be separate individuals, each with a unique conscious awareness inhabiting an empirical reality. We have learned to use our mental and emotional abilities in support of and in resistance to one another, in controlling and in submitting and in just having fun together and inspiring one another. This realm continues to provide us with a great variety of experiences. We now have the opportunity to draw upon all of these and realize the purpose of it all.

We are all energetic expressions of our personhood, acting out our destined roles, until we remember our true identity. We are playing with a wide array of energetics that have been mostly negative in polarity. Those who align with deeply negative thoughts and feelings are mostly out of alignment with our natural conscious life stream. Since they have cut themselves off from the stream of unconditional love that enlivens us all, they need to feed off the energy of those who are fearful, leaving the fearful depleted of life force. In this realm, all negatively-oriented characters become parasites in order to survive without being transformed into their true Selves. They embody the balancing energy for those who are positive.

In order for this world to have legitimacy and believability, it has to have human consciousness providing recognition and energetic alignment, because it is not a divine configuration. In this world, there must be a balance between positive and negative polarities. They require human recognition and constant creation in order to exist. With declining human alignment with them, the polarities begin to dissolve, and we can realize that we are regaining our vitality. We no longer need to feed the negative polarity. The positive melts into a higher dimension that interpenetrates ours and attracts our awareness on a higher-vibratory level.

We can become aware of our conscious Self beyond time and space. We are eternal awareness with person-hood, however we want to express ourselves. If we do not imagine and feel negative energies as if they are ours, they pass through us without taking our life force, and we do not experience them in our lives. Circumstances arrange themselves around the vibratory spectrum of our personal energetic signature, which is our dominant state of being. As we pursue our passions, our world changes with us.

A Path of Inner Guidance

Our nature is pure creativity, and we get to experience everything we create. We can express ourselves in any dimension and in any form or way that we choose. The mental and emotional aspects of our Being vibrate in unison, but at their own intensity. Our vibratory energetic amplitude can differ between thoughts and emotions, because they are in different planes of the intersecting wave patterns of our electrical and magnetic fields.

We have constant free will in every aspect of our lives. We control the focus of our attention and how we feel about it. In order for these feelings to be true, we need to be emotionally free. It can be helpful to be able to realize our eternal presence of awareness as our true reality. This knowing of ourselves can lead to much greater realization within universal consciousness.

We are not required to react to any situation we encounter with any particular emotions or thoughts, but how we react determines the quality of our experience. The vibratory level of our reactions determines the energetic quality of our creative essence, resulting in the quality of what we experience in our personal lives.

By maintaining a perspective of openness, with compassion and joy, we create elevating lives for ourselves, with great radiance of high-vibratory, positive energy for humanity and the Earth. We have a natural knowing of our intuitive guidance. It is available to us by our intent to realize it, through the life-enhancing energy of our heart. We can feel the quality of this energy and align with it, opening us to a state of joyfulness and serenity.

These energetics apply in any dimension. In the world that we experience as humans, we have limited ourselves to a virtuality realm that we believe is real and is negatively polarized with fear, ultimately fear of death and consciousness termination. We're not required to have fear. As part of our human experience, we've acquired it, and we know what it is, and we can intentionally change our focus to positive energies of love. This comes with knowing our eternal presence of awareness.

We exist in many dimensions simultaneously, beyond time and space. Our human embodiment here is one of these dimensions. In the past, we have had to leave the body in order to enter a higher dimension of living. Now we can do this without leaving the body, by raising our conscious vibratory frequency and transforming the physical body to a higher state of being, with the cooperation of our subconscious, innate being.

Knowing Our True Selves

It is possible for us to align ourselves with the greatest love that we can imagine living in and feeling. This is our natural state of being. We took an interesting excursion into dark territory for experience in the negative realm, but we don't have to stay there. The gray zone is a step up, where there is some sense of innate love. When we step into the light completely, love is universal, and gratitude is natural. Everything in our lives is life-enhancing in every way. This is the realm of positive polarity with high vibrations.

The rising resonant frequency patterns of the Earth are indicative of the quality of the larger solar and galactic energetic environment. This shows us that our human resonance is rising as well in order for us to continue to be comfortable while embodied here. There is increasingly more light and love enveloping us. More of us are learning to adjust our consciousness in alignment with higher, positive vibrations.

We can be intentionally aware of any potentiality. It only requires our focus and perhaps our imagination. Whatever we can imagine, we can experience. We have believed that we are limited to empirical experiences, and that life just happens to us. We have lived essentially in reaction to our encounters, having little understanding of what it's all about.

By intentionally searching for the energetic quality of what we want, it arises for us out of universal consciousness through our intuition, which is part of us, even if we aren't aware of it. If we are held back in our energetic creations by our limiting beliefs, we can transcend and resolve them with our intentional state of being, vibrating at the level of gratitude and love.

We are Source Beings, constantly arising out of the consciousness of the Creator of all. We are how the Creator experiences everything we experience, and we thus expand universal consciousness. We are designed to create imaginatively in life-enhancing ways, freely and without limitation. We can love deeply and have as much ecstasy as we desire. In our essence, we are eternal, divine Beings with Self-Awareness, able to create, and allow to manifest, whatever energetic patterns we can imagine. We are infinitely powerful creators, playing in the empirical world as humans. We are learning to trust ourselves, knowing that we are intentionally always in a positive, high-vibratory state of Being.

Managing the Limits of Our Awareness

Through our intention to be open, clear and present, we can feel ourselves being drawn into the consciousness of the Creator in deepest love and greatest joy. This is our natural state of Being, and we are naturally drawn toward this level of vibration. Our attraction comes from deep within our consciousness. While this is happening, we can maintain a perspective of life-enhancing thoughts and feelings in every moment. This is a path to knowing more of our true essence, as we evolve toward full Self-Realization.

Our challenges are greatest at the beginning, because our attempts at openness are blocked by our limiting beliefs. We have not allowed ourselves to realize that we can have infinite awareness. We have not allowed ourselves to open our awareness to realms unknown in our empirical experience or in our ego-consciousness. There is more than one dimension here in our consciousness. Dimensions of potential experiences are present for us to recognize; otherwise, we're not aware of them, and they do not affect us.

For openness to be effective, we can intentionally focus on positive, high-vibratory thoughts and feelings. By transforming our negative experiences into positive ones with our imagination and emotions, we create high-vibratory experiences for ourselves, regardless of the energies held by others around us. This is life-transforming. It requires strong intention and practice to perfect, but the ultimate achievement is beyond wonderful. It elevates everyone around us and brings us into alignment with the rising energetics of Gaia.

Since we are constantly radiating the energetic level of our attention, we can elevate our state of being by paying attention to life-enhancing thoughts and feelings, and by aligning with them intentionally. This kind of presence of Being establishes a positive energy signature, existing in a higher octave of life experiences. Negative influences dissolve into a lower dimension for us.

Chapter 5. Realizing Our Personal Truth

Ego consciousness exists without higher guidance and cannot understand any of this. By intentionally observing ourselves and learning what we actually believe about ourselves, we can decide to make adjustments, if we so desire. We can be open to experiencing eternal present awareness without limits.

A great challenge for us as humans is knowing who we truly are as living, conscious beings. In the empirical world that we share, we intuitively learn to block our awareness beyond our physical senses. Accidentally or intentionally, some of us occasionally have experiences beyond the body, in which we greatly expand our conscious awareness. Those who have experienced near-fatal accidents and medical procedures resulting in temporary death, as well as deep meditations and psychotropic trips have given us reports of realms of pure consciousness beyond the empirical. Here's a link to many reports of after-death experiences: https://www.youtube.com/channel/UC95WPHh1j9fr-BR7gyMOuBeQ/videos.

What happens after physical death? The reports vary greatly, but all agree that we maintain our personal identity as conscious beings. We become conscious awareness without a physical presence. We exist in a dimension that is compatible with our level of vibration and within the confines of our beliefs, just as when we are embodied.

We have been caught in a reincarnation cycle, but this has now been released. Humanity is now in a transition to a higher dimension of living with greater light and joy in our personal being, our relationships and our environment. The polarity of our planet has shifted from negative to positive, from life-diminishing to life-enhancing energies. We are returning to the conscious essence of our Creator. Prompted by the rising positive energetics of Gaia, the Spirit of the Earth, and our enveloping cosmic environment, as partially evidenced on the Shumann Resonance graph (see https://schumann-resonance.earth), we are expanding our consciousness into a higher realm of greater beauty, love and joy.

Physical death is not termination of our consciousness, it is a transition into greater awareness, but without the experiential power of physical life. Our experiences in the body give us deep conscious impressions that are not possible in our spirit presence. As we intentionally move into higher levels of consciousness, our mental and emotional abilities expand, and we become more creative in life-enhancing experiences. This is possible while embodied, and even more so beyond the third-density empirical world.

Realizing Our Personal Truth

Our creative, conscious life force flows to us constantly beyond time and space, arising out of the universal consciousness of the Creator of all. In this eternal state of Being, there is no polarity. Every energy pattern is life-enhancing. Of our free will, we have elected to participate in a realm of polarity. This duality can exist only in a compartment of universal consciousness, limited by time and space.

We have the free will to focus on any energy pattern that we can imagine, loving or fearful, positive or negative, and we have elected to focus our creative attention only in this compartment of consciousness. We have aligned our conscious vibrations with deeply negative vibratory patterns, which we could never take seriously in our unlimited Self. These experiences have given us deeper understanding and compassion, which we have contributed to universal consciousness.

To keep our attention within the confines of the human spectrum of energetics, we imposed upon ourselves beliefs that keep us focused within its parameters. These we can resolve. They are all negatively polarized and are purposely life-diminishing. They limit our freedom, sovereignty, eternal Being, and creative abilities, because these limitations are necessary for a deeply human experience.

If we can take a break from our normal lives for a few moments every now and then, and take a few deep breaths, we can space out and clear our energy. When we can imagine something wonderful happening now, we call forth feelings of joy and love. Just being in this state of being occasionally opens our consciousness toward our expanded Self-Awareness. It is a step toward expanding beyond time and space into full realization of our true Self through our intuitive knowing.

Once we know that we are unlimited Beings, who have chosen to limit ourselves for a purpose, we can participate in that purpose with unlimited awareness.

Awareness of Our Inner Light

We are living in a wondrous time of transformation of all conscious beings from an energetic realm of negative polarity of fear into a realm of positive polarity of joy and beauty. For eons we have been subject to enslavement, torture and suffering so great that it had to be hidden from humanity. Now it is all coming into the light for acceptance and transformation. As this process proceeds, it stimulates chaos, anger, even rage and desire for revenge. There must be balance in justice, but we do not need to be the judges or the enforcers. There are cosmic forces that deal with karma. If we desire to take the path leading to participation in the consciousness of the Creator, we need to be the forgivers and the lovers.

Every living being receives conscious life force from the infinite consciousness of the Creator. We can choose to relate only to this life-enhancing energy and provide no alignment with negative, low-vibration energy. Our state of being is a personal choice, and it expresses itself as our energy signature, which radiates its vibrations around us as our photonic aura. We are the creators in every moment, even if we're unaware of this. We create by our radiant energy, which flows out into the quantum field, which envelops us, and attracts compatibly charged and resonant energetic patterns, which become our experiences.

Although photons express their consciousness in our visible light spectrum, we are aware of them only in large quantities, which we produce when we are fully enlightened. When we are open to receiving the full presence of life force that we are designed to experience, we are capable of carrying and emitting a great charge of energy. We have unlimited electromagnetic energy available for us to create with. In our essence, we are light beings, but in our restricted state of fear, we limit the life-enhancing energy that we can receive by our perspective, our limiting beliefs and our negative thoughts and emotions.

Everything that's happening on our planet now is prompting us to awaken to our true Being, our connection with universal consciousness and our inner light. When we search for it, we can feel the presence of the light in our heart through our intuition, our joy and deepest love. We only need to be open to it and call it into our awareness.

Successfully Transcending Chaos

A Japanese samurai master was once asked by one of his students how he would handle being ambushed by a powerful group of thieves and murderers. His answer: I wouldn't be there. This is how energy works. The master lives in a vibratory level of intuitive knowing and is not subject to assault by negative, lower vibratory scenarios. By our nature, we are the masters of our lives, when we open ourselves to our own higher guidance.

In order to be sensitive to our intuition, we need to be sensitive to positive, high vibratory energy patterns, because this is the energetic spectrum of our intuition. This is the realm of love and joy. It is our eternal Self-awareness. There is no doubt or fear feeding our life force to anything. There is only confidence and knowing. All of this occurs when we shift from negative to positive, from fear to love.

Everything that happens in our lives is a result of our personal creation, with a backdrop of the creations of all of humanity, but the chaos around us is unimportant. When we are heart-centered, our personal experience can be multidimensional. We can live in a positive, high-vibratory state of Being, while also interacting within humanity's vibratory spectrum. When we encounter others, we can recognize and interact only with the light in everyone. We all receive the conscious life force of the Creator constantly, and we decide what to do with it.

Because of our training and programming, we have developed beliefs that keep us from knowing our real abilities. Can we believe that each of us can create universes? In our true Being, we are unlimited. Perhaps we may want to drop all of our beliefs about ourselves and move into the realm of unconditional love and ecstasy in the universal consciousness of the Creator. The human ego cannot fathom any of this, because it is a creature of our limiting beliefs, whom we needed for a convincing experience in the realm of duality. We can love our ego and be grateful for its assistance in this world, but the ego is not us.

Once we open ourselves to Positive polarity completely, we can transcend our limitations through our intuitive knowing. We can befriend our Innate body consciousness and cooperate in physical regeneration and growing vitality. Through our intuition, we can work intentionally with the forces of nature and all of Gaia for the enhancement of life everywhere.

Awareness within Universal Consciousness

We can expand into unconditional love and pure conscious Self-awareness beyond space and time. In our natural state of Being, we are our present Self-awareness within the universal consciousness of the Creator. We are fractals within Creator consciousness, possessing all of the aspects of the Creator. We are eternal Creators. Our ability resides in the conscious vibrations of our thoughts and feelings.

Currently our mental and emotional capabilities are limited by our negative beliefs about ourselves. If we so desire, we can resolve these by working with our intuition in an expanded way. Our intuition is only positive and is part of the life stream that we receive from the consciousness of the Creator. To be receptive to it, we can shift our awareness to the positive emotions around joy and to thoughts aligned with deep wisdom. In this state of Being, we can evaluate our beliefs and resolve them. We have created all of them out of fear of suffering and termination. Everything based in fear has no life force from the Creator. It is only us, who can create fear through choosing a negative polarity for the sake of expanding universal consciousness into the negative and experiencing it, in order to acquire deeper understanding of life.

Once we have decided that we want to expand out of our compartmentalized awareness, we can intend to be positive in every moment. We can ask our guides and angels to augment our intuition with more guidance, and we can then pay attention as much as possible. We can be more observant on mental, emotional and physical levels. We can intend to feel being our higher Self, and we can ask to be drawn into greater awareness of our true Being. It's helpful to do this in a quiet, natural setting, where we can intend to align with the life-enhancing, rising energy of Gaia.

In this process, nothing outside of our own Being is important. We can go deep within, to the center of our Being in great joy and gratitude. As we are able to hold this focus, the conditions and encounters of our lives come into resonance with this spectrum of vibration. Our radiance increases as we reclaim our life force from the negative energetics that we were feeding with our recognition and alignment. We can thus express our energetic presence more strongly around us. We attract others who are in resonance with us, and we make it easier for the newly awakening to expand their awareness into positive, heart-centered energies.

6.

Mastery of life

On the Path Toward Mastery of Life

Imagine being the master of every situation you are in. How do you feel in these situations? We always have the free choice of how we want to feel in any situation. Our emotions hold the magnetic power of our consciousness. Our minds may keep us busy, while our emotions can send the vibrations of our heart throughout our being, if we so intend them to. This center of our consciousness lives in a world vibrating at the frequency of love and vitality. These frequencies arise out of the universal consciousness and provide the life force of our essential Being. We are constantly arising from the quantum field into our personal present consciousness beyond time and space in the universal consciousness of all creation in every present moment for all of eternity. We are destined to know our own presence of Being.

Our move toward conscious expansion can be our intention to be kind and compassionate in every encounter. It doesn't matter what kind of energy we are facing, what matters is our vibration, our mental and emotional radiance. True mastery can occur when we have control of our thoughts and emotions. We can have confidence that we are the creators in every moment, not the reactors. Reactors are creators of the vibrations that they are reacting to. We are always creators with our predominant vibratory frequency. This is the energy stream of the intersection of the planes of our mental and emotional vibratory patterns. It is our state of being in every moment.

We've been accustomed to thinking that we're mortal and that our consciousness terminates with the body. Perhaps this personality may disappear into another dimension, but our present conscious awareness is eternal. We exist beyond time and space. Our personality self is a creative extension of our innermost Being, who is unlimited in every way. We can be aware of anything we want to recognize and energize with our life force as we focus on it in love, gratitude and joy. We can radiate these energies into the quantum field for manifestation in our experience. This becomes easier as the resonant frequency of our planet and our galaxy continues to rise, and it is the direction of the energy flow enveloping humanity.

We are sovereign and free in every moment that we know we are. We can hold this energy in every encounter. Any time we feel fear in any encounter, it's time to adjust our perspective to knowing that each of us is an eternal presence of Being in conscious awareness. We cannot be threatened, because we exist eternally, beyond time and space. Threats are energy patterns that we do not need to align with. Any threat is a parasitic extension attempting to enter our consciousness. We can resolve it into neutrality or even modulate it into a higher-vibrating energy pattern by being firmly our true unconditionally-loving Self.

Understanding Our Situation

We live in a realm that is not supported by the life force of the universal consciousness of the Creator. Our empirical world is constantly created by the conscious recognition of all humans. It is imbued with low-frequency energy patterns of negative polarity. These energetics are life-diminishing, and we can feel this. There is always a tinge of fear, which is not part of our natural state of being. In our essence, we are eternal personal awareness. Fear is impossible, because through our alignment with the resonant vibrations of universal consciousness, we absolutely know that we are our eternal conscious awareness and cannot be threatened. We can be present anywhere we envision.

The entire empirical world has an energy signature, a spectrum that it vibrates within. This spectrum of energy patterns is interpreted by our consciousness as our reality. This is one of the conditions that, prior to our incarnation, we agreed to experience. Our focus on the energy patterns brings them into material form. A photon has an energetic identity, and when we recognize it, it becomes visible light. This is the power of our ability to observe. Observation of energetic patterns results in their changing into our experiences. Subconsciously we expect this to occur. It is the process of creation by envisioning, either through experience or imagination, and feeling ourselves living within the energy spectrum of our thought patterns and emotions. By our recognition and focus on these energetic patterns, within the limits of our expectations and beliefs, we send those vibratory patterns into the quantum field, where they attract compatible energies. When we recognize these, they immediately become our empirical experiences. All of the energies outside of our focus of attention are not changed.

Other dimensions of energetic patterns interpenetrate our realm, and we can become aware of them by raising our vibrations with positive polarity. In our empirical experience, the negative and the positive interpenetrate, and we can choose which one we will adopt as our perspective. In every moment our vibrations attract compatible qualities of experiences.

When we open our awareness to higher realms, we no longer need our beliefs and expectations. When we vibrate on the level of gratitude and love, in alignment with the consciousness of the Creator, we are naturally provided for in abundant ways. Once we can open our awareness to the guidance of our intuition, we begin to participate in universal consciousness. Through our conscious life force in the heart of our Being, we can feel our connection with all naturally-conscious beings and entities.

Understanding and Transcending Our Human Experiences

The Creator of all wishes to create forever, and the life force of the Creator flows to us constantly with the conscious expression of electromagnetic wave patterns, some of which we can perceive and modify as our empirical world. Wishing to expand unlimited consciousness, the Creator imagines us into existence, as extensions of universal consciousness individualized as personal awareness. We are fractals of universal consciousness, able to realize ourselves as infinitely powerful creators with eternally present awareness. Our presence as humans is a small game for us.

All of our experiences here are important. We have become aware of a wide range of duality (positive and negative energies) and vibratory levels, especially negative ones, in a very dense energetic environment, making our experiences intense. Many of these are unique in universal consciousness. They are our mental and emotional transmissions for others to learn from. Throughout the eons, it has taken great courage for us to be here, and we got stuck in the realm of limited consciousness.

When we decide to expand our awareness beyond anything we have known as humans, because we suspect that there's more out there, or rather, deep within ourselves, every experience becomes symbolic of our personal limitations and guidance to the truth of our consciousness. We can be perceptive mentally and emotionally in our constant present awareness. Through our presence of awareness, we can imagine ourselves in a realm of love and joy that actually exists for us. Once we no longer align with fear, we can open ourselves to a higher state of Being and a transformational life.

Without tension, our adrenals become available for greater creative energy in our physical form. We can open ourselves to a greater flow of life force and align with the resonance of Gaia, Spirit of the Earth. By imagining ourselves in a state of quiet ecstasy, we elevate our vibrations in a positive way. By being in this state as much as possible, we eventually transform our entire lives into experiences of gratitude, joy and freedom. We can move between dimensions and remove ourselves from negative polarity, which then does not impinge upon us. Our participation in any dimension requires our consent. We need to open ourselves to it.

If we wish to live in a more desirable dimension, we can be positive and grateful in every way as much as possible in any situation. On an energetic level, we are the creators of our experiences, especially the emotional quality and sometimes the forms as well. We can remember and feel our eternal present awareness, which cannot be threatened or intimidated. Our true awareness encompasses the consciousness of the Creator and includes everything that is the Creator. In our true Self, we are the Creator, fractalized as our personal Selves. We can create universes and more, not in our human-ego compartmentalized consciousness, but in our true, expanded Self.

Enlightening Our Energetic Presence

As we expand our awareness into a mental and emotional void, we can become aware of the quality of energy at the source of our presence of Being. It is our connection with universal consciousness and all conscious life, and it is present in our intuition. It is completely life-enhancing, loving and compassionate. This is the level of vibration that we can mentally and emotionally align with, and we can be adept at calling these feelings and visions into our awareness.

If we have resolved our limiting beliefs and are adept at calling forth mental and emotional states, we gain our natural creative power, undiluted by attachments to negative energy. We can become able to maintain a state of being that is natural and easy, confident in knowing our intuitive guidance in every moment. When we do not engage negative vibrations, we do not align with them or give them our life force. As we align with positive, life-enhancing and creative energies, negative situations disappear from our experience. The problem that most of us face is believing that this is possible.

In our experience, doubt has a negative polarity and cancels out positive, creative energy. The ego must doubt, because it cannot conceive of operating in other dimensions, which is what's in play here. We can live in the realm of duality and at the same time have our state of Being in a realm beyond polarity, a realm of pure unconditional love in creative energy. At this orientation and level of energetics, everything is supportive of joyous being. It cannot be reached by negative energies or beings. Without doubt, negative scenarios that we may face can disappear from our experience into another dimension.

Because our fears and beliefs are hidden deeply within our subconscious, transcending and resolving them may require much practice to accomplish, but this is a necessary step on the way to ultimate freedom. Meanwhile they will continue to attract negative experiences. The more brightness in our state of being, the more enjoyable our experiences become in every dimension that we project ourselves into.

Realizing Our Deepest Love

Because we arise from within the consciousness of the Creator of all, we live eternally in the universal consciousness of unconditional love and joy. As fractals of the Creator, we have the same creative abilities. We are of the same consciousness as the Creator. It is the consciousness of everything that exists. In our expanded consciousness we can create universes. We are as unlimited as the Creator of all.

In order to participate fully in the human experience on Earth, we had to compartmentalize our consciousness, not knowing who we really are and what our true essence is. If we did not limit ourselves with false beliefs of limitation and mortality, we could not have an authentic human experience. We had to program ourselves not to know divine love.

Now everything is changing. We're getting ready to graduate and return to full consciousness. As we can see in the Shumann Resonance Graph, we are receiving massive waves of photons that are illuminating our bodies and our psyches, elevating our vibratory presence. We are being enveloped in universal consciousness, increasing our awareness of the energy of the Creator.

We're able to feel each other's presence more easily. We're becoming able to communicate telepathically with animals and plants. Some of us can recognize the higher-vibration nature spirits. The energetic direction of humanity is toward higher vibrations in our thoughts and emotions. We're learning to transform our self-limitations, heal our psyches and bodies and regenerate our human presence.

As we clear our consciousness of fear and limitation of every kind, we make room for creative love and joy to flow into our awareness and fill our emotional presence. As we open our awareness to greater love enveloping and penetrating us, we transform our lives in compassionate wisdom and begin to feel and experience the unconditional love in the heart of our Creator.

Utilizing Our Creative Essence

We can be the sufficient peacemakers to create peace on Earth. We can reclaim sovereignty for humanity. This is the directional flow of the rising vibrations of the Earth and our surrounding cosmos. Everything is changing in vibrational levels. The world of fear is having its life force withdrawn, as we vibrate in a higher spectrum of energy in which there is no fear. In its higher sense, it is the realm of the eternal present moment in universal consciousness. This is where we can know the true Being of every person in existence, because we all arise from universal consciousness.

In our essential present awareness, we are all participating in the same consciousness. We are the Being Who is always creating everything and everyone. We are in the constant awareness of a higher Being, whose awareness is our conscious life force, and whose essence is unconditional love for all. We are that Being in our personal essence, our expanding Self-awareness. We are our eternal presence of awareness, beyond time and space and embodiment. With our imaginations and emotions, we have the ability to create patterns of energy that become experiences for us.

As we elevate our vibrations, we elevate our life experiences, regardless of what kind of energy we encounter. In the interaction of high- and low-vibrating energies, the high-frequency energies will either attract the low-frequency energies into a higher alignment or they will destabilize the lower frequencies, which will pass into a lower dimension, outside of the higher-frequency spectrum. If we stay in the higher frequencies, we encounter only high-frequency experiences in our Being. In recognizing low-frequency energy patterns, we can transform them by sending them our acceptance, compassion, and love from the perspective of compassionate understanding.

We control our lives by how and where we send our life force and how we interact with the energies we have sent out by our thoughts and feelings. Our energetic signature radiates out photons that express our energy. The vibratory radiance of these photons attracts energy patterns that align with us. This is how we feel the presence of each other and know when our vibrations are in alignment.

By aligning with peace and love, we send our life force into these vibrations and empower them to radiate out to humanity and the cosmos. These energies are in alignment with the conscious intent of the Creator, whose vibrations are always life-enhancing. By using our consciousness in this way, we greatly empower the energies of peace and love among humanity and our planet.

Staying in the Love Vibration

When we become concerned and angry about our contrived human condition, we will continue to give our life force to the conditions and people that we oppose through our energetic alignment with them. We may realize that we are always expressing the vibratory quality of our thoughts and emotions. This expression creates our personal energy signature, which is our predominant frequency that creates the quality of our experiences.

Unknowingly, we are the masters of our lives. If we think of life as continuous encounters with energy patterns of all kinds, we can choose which ones to recognize and focus on. We can understand how we create our reality by radiating our energy signatures into the quantum field, where we attract energy patterns that align with our vibration and repel those that don't. Our emotions are magnetic with our own polarity, in combination with our imagination's electrical polarities. We attract and are attracted to resonant energy patterns. In our empirical world, these energy patterns come into our recognition as material experiences. Those innumerable energy patterns, that we do not recognize in the quantum field, do not become part of our experience.

Our bodies radiate our electromagnetic frequencies and polarities. By paying attention to our subtle, deep feelings, we can know everything we need to know in each moment. This comes by being aware of the frequencies of unconditional love, compassion, forgiveness and gratitude. By paying attention to those feelings, we can use our imagination to elicit high-vibration experiences. Our personal vibratory spectrum attracts and manifests the quality of our experiences. By raising our energetic frequencies, we become capable of creating miracles.

Once we realize our mastery of energetics as a spiritual practice, we can discipline ourselves to hold a high-frequency perspective in all encounters. This transforms any low-vibration energy in our experiences into alignment with us, or it disappears from our experience. Once our focus stays in the love vibration, that is what we are continuously creating for our experiences.

Interacting within Humanity's Energetic Spectrum

It's a challenge for us to know that anything is possible, and that we can live miraculous lives, regardless of what has been happening in our awareness and all around us. This is the leap in consciousness that can greatly enhance our experiences on this planet and beyond. It requires us to stop aligning our thoughts and emotions with the low-vibrational energies that appear to threaten and intimidate us with destructive, parasitic intent. When we are angered and fight against these low-vibratory energy patterns, we give them our life force, because we are meeting them at their own vibratory level. Without our consent for engagement, they cannot invade our consciousness. This is where miracles can happen.

When we have cleared our consciousness of our limited beliefs about ourselves, we can open ourselves to our true essence, our eternal Self. We can do this now. The only powers holding us back from expanding our consciousness are our own beliefs that limit us. These are beliefs such as thinking that others can make us suffer, starve us and imprison us. We believe that we are inherently sinful, shameful, untrustworthy, and capable of growing old and frail and getting sick and dying. These are all false beliefs that we have acquired and made real for ourselves. They are all part of the human hypnotic trance. It is a challenge to resolve them into nothingness, but it is possible with great intent.

There is no proof that we will be successful in this quest. We have been taught that the great masters, all of whom have accomplished this transcendent state of being, are different from us, even though they have told us that they are just like us. Jesus told us that we could do everything that he could do and more. Now we know how this can happen.

In the perspective of quantum energetics, our entire life experience is an expression of different frequencies of energetic vibrations, which we control with our thoughts and feelings. The quantum field of all potentialities consists of conscious life force flowing in unconditional love. We arise out of the universal consciousness of the Creator of all and exist in the energetic plasma of the quantum field, where everything is possible. We are limited only by ourselves in expressing anything that our heart desires. This is the other requirement for a miraculous life. We are designed to express the energies of the heart of our Being in order to live in the spectrum of high vibrations of our fulfillment in joy, love, freedom and abundance, guided by our intuition.

When we maintain a conscious intent of compassion, gratitude and love, we remove ourselves from the realm of low vibrations. As long as we're participating in the human hypnotic trance, we'll have to face low-vibration energies, and we can interact from the perspective of compassionate wisdom, removing ourselves from alignment with the dark force and its consequences.

Living Multi-Dimensionally

Many of us expect our experiences to deteriorate further into the lower vibrations of suffering and evil. Fascism and division of many kinds are being promoted in the media. The world we have now is a result of the vibrations that humanity has created or allowed to be created with our life force. We live in a world of duality, and we have a choice of what level of vibrations we prefer to experience individually and collectively. This is a lesson that all must learn.

As multi-dimensional beings, we can live in different worlds of experience without going anywhere physically. The dimensions all exist concurrently and occupy the same space in our consciousness, just at different levels of frequencies. The qualities of our experiences differ greatly from one to another. In the lower frequencies of our thoughts and emotions, we experience fear and deterioration of our being. In the higher frequencies we experience life enhancement. We have absolute control over which quality of life we choose to experience.

As we choose and intend to live in a high-vibrational life, we elevate our personal energy signatures. These radiate our higher-vibrational consciousness into the life-manifesting unified quantum field. Enveloped in the etheric plasma, we can recognize an infinite combination of energy patterns that we can attract into what we wish to experience.

As humans living within our limited sphere of conscious awareness, we have chosen to keep our awareness limited to the empirical spectrum of energy. But there is much more that we can be aware of. We are our only limitation from universal consciousness. It has been our belief that we are identical to our bodies, and we are mortal. Once we have crossed over in our awareness to a higher dimension, we know the truth of who we are as personal eternal Beings. But without making this leap in consciousness, we can acquire an understanding of quantum energetics.

By knowing how we experience different vibrations, we can know that we can create whatever quality of vibration that we can imagine and feel ourselves living in. This is where our limitations come in. We can't imagine the reality of something that we don't believe. False beliefs about ourselves keep us from realizing our eternal Being. As long as we believe ourselves to be mortal, it's a leap in consciousness to know that we are eternal, that we are conscious persons who are expressing ourselves through our physical presence in the body, and we have a personal conscious awareness that is unlimited in every way.

As we become comfortable in awareness beyond the body, we can more easily create the kinds of energies we want to live in and then manifest these electromagnetic wave patterns into physical experience. This is when everything becomes magical in our empirical experiences.

Creating Out of Pure Awareness

We are designed as Creators. Our consciousness is part of the universal One. Quantum physics has shown that there is only One universal consciousness. It is the consciousness of the Creator of all that exists.

In our current form in the dimension of human experience on this planet, we have existed in a compartmentalized consciousness that allows us to use our free will to experience fear and suffering in order to deepen our understanding of love and compassion. In our true essence of eternal unconditional love and infinitely powerful creativity, we could never experience the lower vibrations that we have experienced here. In order to make this experience as real as possible, we have created our ego consciousness with a rational mind that has no awareness of our true Being.

We have learned to make plans and think about how we want to live and what we want to experience. We've learned to endure hardship, not knowing that we can transform our lives, but this is the experience we planned for ourselves. There have always been incarnated masters to show us the way back to our full conscious awareness, and those who were ready to return have paid attention to them and learned how to return to our eternal Being. There have also been those who came to rule over us and keep us in limited consciousness through fear, and they distorted the truths that the masters taught in order to enslave us.

Now we are at the turning of the ages, and our planet is rising in conscious vibrations. In order to continue to live here, we must release our focus on low-vibratory experiences and situations, transforming ourselves into our higher-dimensional Selves. This process requires us to stop recognizing and focusing on our limitations and all of the evil that still exists here. We are being prompted to open our awareness to the energy at the heart of our Being. This is where we find our conscious, unconditionally loving life force flowing into us from the essence of our Creator. We can find this source of our life in deep meditation and contemplation. It requires practice and feels amazingly wonderful.

Once we become aware of our inner truth of Being, we are on the path to expanded consciousness and unlimited creative ability. This is a state of Self-Realization of present awareness in divine love and joy. In this awareness, everything is possible. All of our needs are effortlessly fulfilled, and we are not limited to the consciousness of physicality. We have the opportunity to be our true, unlimited, deeply loving and compassionate, naturally creative, expanding Selves. We can be our own angelic Presence.

Greater Levels of Realization

We can continue to go deeper into our conscious self-awareness by intentionally increasing our feelings toward the purist love that we can imagine. By paying attention to these feelings and the visions that accompany them, we can raise our vibrational frequency into closer alignment with our Creator Self. The Creator expression of our Source of Being constantly streams our conscious, unconditionally loving, life force into our awareness. We perceive as much of it as we allow ourselves

Our false beliefs about ourselves keep us from realizing our unlimited Self. Considering our potential, why do any of us continue to keep our limiting perspective? We have become so captivated by the spectrum of energy that humanity vibrates within, that we do not realize that we've been in a kind of hypnotic trance, keeping us unaware that the consciousness that we live within is only a tiny part of who we are. We exist beyond time and space as pure, unlimited Self-Awareness, knowing how to create everything we imagine and feel, enabling us to live in a scenario of our focused vibratory level.

Once we get the idea that we can be more than we have been, we can begin to open ourselves to greater love. This must be an emotionally intentional move in order to align our subconscious being with our intentional vibratory spectrum. It is a direct way of resolving our limitations. In order to love our entire Being, we must intend to be accepting, forgiving and loving, while realizing the inner light of the Creator in every encounter. This is the level of vibration that we can elicit in alignment with our greater love and joy.

All of our life experiences can become acts of love, out of the universal consciousness of the Creator. When we intentionally recognize energy patterns that feel loving, we begin to live in this vibration. The deeper we can go, we can approach the greatest love of All. We can focus on this level of vibration in our realization of our unlimited Self-awareness. We begin to realize our multidimensional Self, able to manifest ourselves in more than one dimension simultaneously. This is the beginning of a new way of life, inspired constantly in gratitude and joy, aligned with the energies of the Spirit of the Earth and all life-enhancing perspectives.

Awareness of Our Energetic Balance

Each of us is a unique Person, capable of knowing the expanse of universal consciousness and the unconditional love that constantly flows to us in the life force of the consciousness of the Creator of all. In our natural Being, we can realize the magnitude of our universal consciousness, but in our compartmentalized consciousness within the spectrum of vibrations of humanity, we have closed ourselves off from knowing the Creator. We do this by ignoring our intuition and conscience. We have trained our subconscious minds to filter out the messages from our heart, but if we feel drawn to the heart of our Being, we become aware of our innate Being beyond the energetic dimension of humanity's vibrational perception. Our innate Being is aligned with joy and love, but is willing to follow our choices in serving us. As we delve into the realm of fear, our innate Being empowers us to manifest fearful situations for us to experience. It helps us in whatever way we choose to go.

We have the ability to choose to exist and be in the high vibrational realm of love and joy. It is our natural state of Being. It is our pure present awareness, observing all and aligning with the vibrations that we choose to focus on. The focus of our attention and emotional awareness is much more than just our thoughts and feelings that run through us all the time. It is our creative stream of Being. The vibratory level that we choose to inhabit with our awareness provides the energy signature that attracts experiences that are in alignment with us.

Becoming aware of our energetic balance can allow us to know what we're constantly creating. This can be intentional for us. We can jump around in our mental and emotional polarity and between high and low frequencies. If we choose positive, high-frequency vibratory patterns to recognize and align with, we can come into the presence of our true Self and the realm of greatest love, joy and eternal presence of awareness. We can know how we are multidimensional and can have universal consciousness in the dimension of love and gratitude. We can also have limited consciousness in the dimension of humanity's vibrations. The power and free choice of how we use our attention determines our energetic expression, which radiates around our physical presence in the form of conscious photons that we emit constantly. These come from our life force through our attention, which is why we can see our vibratory level in our eyes. Photons are conscious, subatomic quantum light beings. The ones that we emit are in alignment with our energetic expression and are colored as such in our aura.

Being Aware of Our Intuitive Guidance

By being sensitive to our intuition, we can be able to know everything we need to know in every moment. We may not have yet been able to know what intuition is or how to be aware of it. It arises from the heart of our Being. It is what we truly know in our deepest knowing. It is not influenced by anything outside of itself. It is moved by unconditional love and compassion for us. It is a constant presence beyond the conscious ego mind. It is part of our eternal awareness and is how we can know the unknowable.

We can become sensitive to the promptings and knowing of our intuition by being clear in our intentions and not wanting anything. Intuition is beyond words and can only be alluded to. By being impartial observers of ourselves, we can be aware of our intuitive guidance in what we need to know. We can just know what to think and do.

Because we are accustomed to relying upon our conscious ego mind, we have learned to tune out our intuition. We have relied on cognition primarily to navigate life. Now we can adjust our perspective to be open to greater awareness that comes from higher vibrations of Being. By imagining ourselves living in a high-vibratory environment, we can enjoy encounters with kind and loving people and find ourselves in beautiful and magnificent environments. When we feel guidance toward life-enhancing thoughts and feelings, it comes from our intuition. This is our natural way of knowing about life.

In order to be our natural Selves, we must abandon feelings of all low-vibration states of being and replace them with what feels good and loving. Our imagination can be unlimited and can jump right over our limiting beliefs about ourselves, although we do need to resolve these, and it's easier from an expanding perspective.

We can learn to be sensitive to our intuition in a precise way, so that we always know that we are properly guided in everything, and we have the free will to make all decisions about our life. Realizing the love in our heart and our intuitive guidance is all we need to master our lives and raise the vibrations of humanity.

Moving Along the Inner Journey

As we move further along in our inward journey toward the light of our Being, our ego consciousness responds positively to our love. Our ego just needs to know that we value and love ourselves, and we care for our well-being. We have had to begin our journey from within our ego consciousness, which has not known anything beyond its own level of vibrations. The ego has self-awareness, but it does not know who it really is, because it was designed to operate within the consciousness compartment of time and space.

When we decide to lift the curtain and look beyond our self-imposed limitations, we begin to enter the quantum realm of universal consciousness, and our awareness becomes unlimited. Here we transcend the ego by opening our awareness into our expanded personhood. We can imagine more wonderful circumstances for us to live in. We can feel ourselves living in love, compassion, joy, abundance, gratitude and freedom. These can be the vibrations we are creating for ourselves to experience. The more we do this, the more real it becomes, until we realize that we are really living in our envisioned and felt quality of experiences.

With practice we can become proficient in staying in high-vibration, positive thoughts and emotions, regardless of what kind of energy we are facing. We learn to define our own roles in life's drama. We get to experience the greatest love and joy. This happens when we realize that our present awareness is beyond time and space. We are eternal Beings beyond the empirical world of negatively-polarized, low-vibration experiences.

All we really need to do to become masters of our situation in every moment is to stay in high vibrations. By focusing on the higher octaves of energetic patterns in every situation, we can recognize and feel the presence of Creator consciousness, the life force and inner light that constantly enlivens us and connects us with all conscious beings. We can accept our eternal nature and realize that we are the Creator in expression of our individual personhood. Through our intuition we receive the conscious guidance of unlimited awareness. We are designed to be sensitive to our inner knowing. It only requires our attention.

Experience Beauty, Love and all of their Associates

The only limitations to our full knowing of our own true Being, are self-imposed. They are borders in our awareness that deny our access to universal consciousness. We have allowed ourselves to believe that we are mortal ego-consciousness, separate from our Creator. We needed this for our deepest, most meaningful human experiences in the negative, low-frequency environment of the life-diminishing world of humanity. But there is more to us, much more.

We have become masters of darkness. It has cost us most of

our life force, because our limitations shut down our positive, unconditionally loving conscious life stream. We've learned how to divert life force from others, attempting to make up for our lack. We have survived by the strength of our ego consciousness without higher guidance. We have been mostly unaware of our intuitive guidance in every moment, and haven't even known that it exists.

It does exist, very faintly, if it has been mostly ignored, but we can become aware of it, if we intentionally open ourselves to our inner knowing. It is the Source of our consciousness and our vitality. It is who we are as extensions of Divine Being and is our connection to universal consciousness, the consciousness of the Creator. It is our constant guidance in the level of our vibrations. It allows us to create any kind of life that we choose to focus upon by our conscious alignment with its energetic expression. It is the quality of energetics that we create for our experience with our vibrations.

Our intuition can always help to keep us aligned with higher consciousness. Our intuitive guidance is not normally given to us in words. It is symbolic, and it prompts us in ways that it knows we can understand. We just have to be unattached to everyone and everything, and just be present in awareness. In this moment we can receive a knowing of whatever we need to feel or imagine in our current situation. This may motivate us to say or do something. Sometimes it may not make any sense, but the more we become sensitive to our inner knowing, that more accurate we become, and the more we can know that the beliefs we have held about ourselves are no longer needed. Then we can begin to have access into our expanded consciousness and mastery of life in any conscious realm.

Our Higher Guidance

When we are ready to understand our empirical world from a

higher perspective, we can expand our awareness beyond time and space and into our eternal presence of self-aware Being. If we seek to know the Source of our conscious vitality, we will find it, because it is always present in our intuition.

We can recognize the feeling of energetic polarity patterns. We can feel negative, life-diminishing frequency patterns as well as positive, life-enhancing encounters. We can enjoy the positive ones, which bring joy, and we can transform the negative ones, which solicit fear, by maintaining awareness of our eternal Being, which is positive and emitting high-frequency radiance.

There are many ways to mastery of life. At the basis of all of them is the realization of our Source-consciousness. We can open our awareness into the infinite conscious knowing of the Creator of all conscious beings. We arise out of and are the One consciousness. In our true Being, we have infinite awareness and infinite creative power.

Currently we're practicing raising our vibrations to a higher dimension. We can do this at our own pace.

At some point, our inner journey can become the most important focus of our life. From here on, no written teachings or videos really help us. We enter the unknown realm of a higher dimension of energetics. Everything works differently. People are kind and caring. There is unanimity with everyone. We can understand that there is no good and bad. It's all a consciously-confined play, and we've delved deeply into our roles.

We can know that we're living beyond the physical world. We're all the same Being of light and love. We can feel the inner light of everyone. As we develop great sensitivity to our intuition, we can be completely guided in every moment, always knowing the highest frequency vibratory patterns to engage and align with. We'll find that circumstances will arrange themselves to provide experiences in alignment with our own energetic signature. We can realize that we are the energetic modulators and creators of all aspects of our lives.

Continuing to Expand Our Awareness

On the inner path toward being our true Self, we can seek the vibrations of love and compassion. These are expressions from the heart of our Being and are radiating within our perceived being and all around us. We've been distracted from inner knowing, by being constantly programmed to live in ego-consciousness. We have been unaware of our inner knowing, while it is the source of truly knowing anything.

We can keep searching for higher vibratory patterns to come into our awareness. As we encounter others, there are predominant vibrations that we all carry in our consciousness. When we are aware of how we feel in each encounter, we instantly know the polarity and frequency level of the vibrations. If we feel some negativity, we can recognize it and envelop it in our positive radiance of Being. If we can find some light in the beings we encounter, we can relate to that light--that glimmer of Creator Consciousness. If there is no positive connection, we can still maintain our own polarity and frequency patterns. Situations then arrange themselves for the energy to come into alignment with the quality of our focused attention.

As we gain proficiency in maintaining a focus on a possibly real, possibly imaginary, world that is wonderful in all ways, we give it more of our life force through emotional knowing, we begin to make the transition into a higher dimension of energetics and living. As we delve into it, we begin to feel greater joy and gratitude for the wonders of our conscious awareness, which can be completely clear. We can begin to feel much more deeply, with more nuances in our emotional awareness.

On the way to clarity, we can learn to become aware of our awareness. We can become consciousness watching consciousness. As we become aware of the qualities of energies in our awareness, we can intentionally seek alignment with high-vibration feelings and life-enhancing thoughts. When we can do this in our encounters, regardless of what the situation may be, we can become masters at modulating energy into alignment with the life-creating energies of Creator Consciousness. We can begin to enter into our eternal, infinite awareness.

Realizing Enlightenment

If we decide to practice feeling the most expansive, joyful and ecstatic scenarios we can imagine, we begin to expand our awareness beyond the physical. If we continue this practice, we can align our entire consciousness with these vibrations. Then we'll be able to realize their reality in our lives. Recognition of higher vibratory energy patterns is an expansion of our awareness that we can intentionally achieve.

To open our awareness to a higher dimension of mental and emotional frequencies, we can seek them and ask them to come into our awareness. Intuitively, we know them, and we know how they feel. They radiate the energy of our eternal Being. If we search for them with a clear heart, we can realize them. We can feel their positive magnetic polarity, which we are naturally attracted to. This is a realm of beauty and abundance. We can get here by focusing on scenarios that express this spectrum of vibrations.

As we go deeper into our consciousness, we can realize that we are each an eternal presence of awareness. Currently we are projecting our consciousness into our physical presence as humans. Although we have been aware of the spectrum of empiricism, we have not known about the consciousness that creates the material. That consciousness is us. We are the creators of this entire realm of experience, and we have the potential of being the masters of it and so much more.

Our entire lives are created in the focus of our awareness by the vibratory quality of our attention. We are constantly modulating and creating our energy signatures with our changing mental and emotional states. If we can keep our awareness of our intuitive knowing, we can live in alignment with life-enhancing experiences, independently of the negative realm of humanity. The deeper we go into our intuitive knowing, the more expansive our awareness becomes, and the greater our life experiences.

Transforming Negative, Fearful Situations

Incarnation on this planet as humans has given us the opportunity to enjoy the pleasure of embodiment as well as learning to rise to the challenge of being stuck in fear, suffering and pain. The challenge is learning how to extract ourselves from negative polarity in our thoughts and emotions and then remain free to experience only the vibratory level of love and joy. Every moment offers us a choice of how we want to experience it. The more proficient we become at choosing the positive perspective, the more dramatic our challenges can become, until we finally arrive at mastery of ourselves.

As humans, we have accepted, and imposed upon ourselves, beliefs about our limitations. We no longer need these, once we decide to embark on the inner journey to knowing our true Self. When we are aware of a limitation, we can resolve it through acceptance and forgiveness of ourselves for this intentional imposition on our consciousness. We can love ourselves for wanting to expand our understanding and compassion through negative, life-diminishing experiences that seem real.

We have the opportunity of encountering our challenges in many ways. Some of us learn to toughen ourselves to be able to fight and endure extreme pain without breaking our focus. This approach teaches mastery of the focus of our attention, but it does not leave the realm of the negative. Some of us learn to meditate deeply or use psychotropic substances enough to transcend the awareness limited to the physical body. We may travel in the astral realm, and here we can meet beings much more powerful and fearful than in the physical world, if we are still focused with a negative polarity. The quality of what we recognize and encounter with our focus, becomes manifest in our experience.

Ultimately we can learn to focus only on what we prefer. We can be aware of a great spectrum of energies, and we can choose which ones we will engage with and align with. When we consistently choose to understand life through a perspective based on a positive polarity with high-frequency emotions and thoughts, we transform our ongoing situations into experiences of freedom and joy. We develop the ability to focus on life-enhancing situations and outcomes and to withdraw our life force from all negative, life-diminishing energies and beings. We can choose to disengage with them in our focus, or we can encounter them while maintaining a positive, high-vibratory perspective. This process requires intentional practice and determination. It can enable us to live in the universal consciousness of the Creator of all, which is our natural state of Being, and in which we eternally exist beyond the body.

Once we have resolved all of our self-imposed limitations to our conscious awareness, we can open ourselves to our eternal presence of awareness at any time, while also being aware of our awareness in the body. We can become aware of our deeper intentions in each experience.

The Presence of One

Creator Consciousness operates in every moment, issuing life and form to everything in existence. We are part of this creative consciousness. We are constantly creating energy patterns of the vibratory frequency of our mental and emotional state of being. By recognizing these frequency patterns and aligning with them, our innate consciousness interprets them as empirical for us, and we bring them into our experience. As we recognize electromagnetic wave patterns, we can modulate their frequency and amplitude by what we think and feel about them.

If we have no inhibitions about our abilities, and we can be fully present in our awareness, we can enter a dimension of vibrations much greater than the empirical. It controls the empirical from a quantum level that is beyond time and space. Our presence of awareness is the most powerful aspect of our Being. It is our womb of creation. It is what we share in unity with all conscious beings. If we desire, we can open our awareness to the consciousness of the Creator of all. That consciousness is constantly enlivening and forming us according to our free-will state of being.

We are completely self-determined, and we have the potential to realize a wonderful state of Being at any time we choose. This must be intentional, because it requires motivation to maintain a positive polarity with high vibrations. When we are fully present in our awareness, we can feel joyful and good in every way. We gain the ability to change the quality of our circumstances and experiences. We can work intentionally with universal consciousness to create a wonderful energetic environment for ourselves and invite others to share it with us. This energy is contagious and can expand throughout humanity.

For eons we have believed that we are separate embodied individuals, but this is just one of our expressions in a limited spectrum of energy. We have believed that our ego consciousness is our complete being, but any time we desire, we can open our awareness to our eternal Being, the One consciousness that constantly creates everything. Through our intuition, we can feel and know this Presence of Awareness, which constantly comes to us throughout our entire Being. In every moment, we can align with this level of energy in absolute confidence and knowing.

How Deep is Our Being?

In the inward journey, we encounter our innate being, the master of our physical consciousness and the repository of our ancestral energetic inheritance. Our subconscious, innate being is a compartmentalized consciousness within our greater Being. It is designed to follow our guidance. It is aware of the vibratory frequency of our ego-consciousness, while also having to align with our ancestral energetics.

We are designed to direct our lives through our intentional alignment with positive, high-frequency patterns of energy. These stimulate our emotions with feelings of joy and gratitude. We can intentionally choose to feel this level of vibrancy. Our physical bodies are designed to be beautiful and vibrantly healthy.

Our innate being cannot judge us or punish us. Any defects are a result of our alignment with low-frequency states of being. This kind of energy is what we inherited and can resolve through compassionate wisdom. As we become aware of inner negativity, which we can do, if we want to, we can resolve it, especially if we are sensitive to our intuition, and even if it seems to be outside of us.

We can intentionally communicate with our innate being, who is aware of our awareness, and whose awareness we can also be aware of. Our innate doesn't understand as we do, but it knows our level of vibrations. As we become more joyful and thoughtful, any anomalous energy that our innate is holding becomes obvious, so that we may accept it with forgiveness and compassion. It's all part of our inherited experience, which has brought us to our present state of being. We are alive and can be expanding our conscious awareness in gratitude.

We can accelerate the process of Self-Realization by intentionally resolving our self-imposed and inherited, limiting beliefs, while also choosing to pay attention to high-vibration energy scenarios. We can choose to imagine beauty in what we see. We can imagine kindness in everyone we encounter. If we look deeply, we can find it, except in the unrepairable people, who are leaving.

We can penetrate deeper. As we begin to realize our eternal presence of awareness, we can become aware of the awareness of all conscious beings. We are all constantly arising from the universal consciousness of the Creator. We are our free-will awareness, attracting our thoughts and emotions. With practice, we can ultimately elevate and expand our awareness into infinite Being with infinite abilities. Each of us is the Creator. We are fractals of the whole.

Realizing Greater Awareness

Although in our true Being, we have unlimited abilities, we have kept ourselves as humans from enjoying our true creativity. This is the situation of our ego consciousness, living without higher guidance. Our consciousness has been anchored in negative polarity for eons, but we are not required to stay there. We can change at any time by our personal choice. Belief in our mortality holds us back, as do beliefs that we need things from others, that we are subject to governmental control and many other limiting beliefs.

Once we get an intuitive prompt that we have greater potential, we can search for more. At some point we realize that experience within the energy spectrum of humanity is a kind of augmented reality that we've agreed to participate in for our spiritual growth. We intended to increase our awareness of the feelings and thoughts in the realm of negative magnetic energetics. We needed limitations in order to experience this spectrum of energies as real. We're in a 3-dimensional movie that we have written, directed and produced for ourselves, but we have been unaware of this, and not knowing the greater design.

To find out about this, we can awaken our connection with our subconscious innate being. It is the part of our consciousness that interprets our intentions into the electromagnetic wave patterns in the quantum field, and it remembers every detail of all of our lifetimes. By using kinesiology techniques, we can ask our innate specific questions that do not require any reasoning ability, only straight answers, such as yes or no. As we progress, we can begin to be aware naturally of what our innate can say about anything. We can also develop sensitivity to our intuition. We are designed to be the director of our innate being, but we have allowed our ego-consciousness to operate freely, and so the ego has trained our innate consciousness, resulting in much negative chaos within, keeping ourselves from greater awareness.

If we have a strong intent to master ourselves, we naturally create awareness of a way that works for each of us. It may even require miraculous-appearing interventions, but we do succeed, because it is in the current flow of life on this planet, and circumstances arrange themselves to accommodate our intentions. Our intent is how we recognize our true eternal Being without limits. This is the inner path to Self-Realization.

Experiencing the World We Desire

High vibrations exist in every circumstance. They are part of the conscious life force that expresses itself through vibratory patterns of electromagnetic waves. They are waiting for us to recognize them and realize their fullness. We can become aware of these wave patterns by intending to recognize and feel their presence. They are always present, and we can open ourselves to them.

All dimensions are present for us to experience, and we can choose which one we want to live in. Currently we have chosen to participate in the empirical realm and to live within its scope of vibrations. In order to do this, we've needed limiting beliefs. Whenever we feel that we have thoroughly experienced the thought patterns and emotions of this realm, we can choose to live in a higher-vibratory realm. Instead of completely aligning with the energies that we encounter, we can choose to expand into the dimension of joy, abundance and fulfillment. By imagining and feeling the higher positive energies that are present in our experiences, we can align with our intuitive knowing.

We have the ability to re-polarize negative energies in our personal experiences by having an intentionally positive perspective. Since consciousness is the source of everything we experience, our consciousness creates the quality of our experiences by our vibratory levels. By being in gratitude and joy, we can transform our lives. By aligning our thoughts and feelings with love, joy and compassion, we can access the feelings of our higher Selves.

In the higher dimension that we are evolving into through the expanding energy of the heart of our Being, there is only a positive polarity of love, joy and abundance. Fear is absent. We no longer need its lessons. We cannot be threatened, because we are the commanding consciousness. We are eternally sovereign, and this realization gives us mastery of the empirical world. All of our limiting beliefs can fall away, if we recognize their essence as life-diminishing.

If we choose to feel and know high, positive vibrations, we can make this our predominant energy signature. In this perspective, we can no longer be impoverished or controlled. We can be grateful that we live in a world of great beauty and majesty, and that we can live in harmony and joy with Gaia and with one another and all creatures. As we learn to recognize and realize this world, it becomes real for us.

Realizing Our Personal Freedom and Sovereignty

The quality of our personal lives depends upon the vibratory polarity and frequency of our own state of being. It does not depend upon our actions, except that they are an expression of our personal vibration. We are involved in a dramatic holographic kind of play in which we are characters and actors with predetermined events and experiences. What isn't predetermined are our intentions, emotions, and perspective. These are the creators of the quality of our experience within the empirical play.

We have absolute freedom to think and feel whatever and however we choose to focus our attention on and align our vibrations with. There may be great distractions going on around us. There could be war, fires, starvation, and disease. These are only distractions from our inner journey to expand our awareness into our true, greater Being. When faced with great distractions, we need great resolve to be on the inward journey. We never stop facing challenges. This is how we expand our awareness.

Somewhere along the way, we make many realizations of our greater reality. We become aware of our eternal presence of awareness beyond the body. Our physical presence is a limited expression of our consciousness. We learn that we can cast our emotions forth into whatever we want to feel and project. We can create visions of wonder and amazement. We can realize that we already have absolute freedom in our eternal Being. Once we know that we are our personal eternal conscious awareness, we can recognize that everything responds to our conscious vibrations.

In our empirical personal drama, predestined circumstances always arrange themselves for us personally in a variable way in relation to everything around us, and in the vibratory spectrum of our state of being, constantly and in every moment. When we are in a state of compassion, gratitude and joy, something can always happen to protect us and enhance our lives, even though we're living through the human drama. We become impervious to threats, because we know how to follow our intuitive knowing to adjust the vibratory level of our experience. The more of us who choose to live in freedom and sovereignty, the more the human drama is elevated.

Going Deeper into Inner Knowing

The masters of life on this planet, those who live miraculous lives, have always known that consciousness is the source of everything that exists. Consciousness is One. It is universal in everything, and we can be aware of the awareness of its entirety. Some quantum physicists have come to a similar conclusion, which brings science and spirituality into alignment. In order to do so, we can resolve our limitations and open ourselves to the consciousness of the Creator. This is the awareness we are designed to participate in. It is our natural state of Being.

By resolving our limiting beliefs about ourselves, we increase our vitality, our life force. By withdrawing our attention from any distractions, we can focus on our own state of being. We can change our vibrational alignment from negative to positive in every moment. We can accept how we feel about ourselves and be compassionate and intentionally loving and understanding about ourselves in the deepest ways, beyond ego consciousness. We can come into alignment with the energy of the heart of our Being by adjusting our own vibratory level to the level of gratitude and love. These are our natural emotions, when we are aware of ourselves beyond time and space.

Along the way we encounter our deepest fear, that of extreme torture and termination of consciousness, but consciousness cannot terminate. It is everywhere always. We are part of it. We can know deeply that our personal awareness is eternal. There is no other possibility. All of our beliefs about ourselves have been shown to be false. We are unlimited in our true awareness.

As we learn to control our focus of attention, while being aware of our emotions, we can be masters of this realm of limited consciousness that humans inhabit. We can become true in every way, and we can live in joy and deep understanding.

When we can completely trust ourselves in every moment, we gain our full life force and awareness into universal consciousness. This awareness is available to us always, whenever we can align ourselves with its polarity and vibratory level. It is the energy of our true heart.

Developing Deep Understanding

It is possible for us to realize the pure conscious connection with the Source of our Being. Through our inner search for the feeling of unconditional love and life-enhancing creativity that is constantly present in the inspiration arising from the heart of our being, we can be present in transparent awareness. We can become fully conscious of our eternal Self, our always present awareness without limitation, having infinite creative power.

Our ego-consciousness cannot fathom this, because it is bound by its nature to believe in limitation. Our unlimited Being is beyond our conception, yet, if we go deep enough into our intuitive knowing, we can understand that it is true. In our human roles, we are players in a game that we designed and directed to give us realistic experiences to expand our consciousness into the negative polarity of fear, and then to learn how to realize that we have confined ourselves within a compartment of consciousness, limited by our beliefs about ourselves.

When we choose to explore the limits of our consciousness in intentional ways, we can examine our beliefs. Do our beliefs benefit us? Do they contribute to the enhancement of our lives? If it is possible to be unlimited, why should we choose to be limited? Resolution is possible through our intuition. If we can align ourselves with feelings of freedom, joy, and gratitude, we can become sensitive to our intuitive knowing, which arises out of the conscious life force that we constantly receive from universal consciousness.

When we can change our focus from negative energy patterns, that are limiting and life-diminishing, to positive, unlimited and life-enhancing energetic scenarios, we become masters of our human lives, because we can understand the game and be guided by our intuitive knowing. There is nothing keeping us from full realization of our true Being, except our self-imposed beliefs and with no desire to expand our awareness into the consciousness of the Creator, which we are created to do and be. Beyond our humanity we are wondrous Beings, and we can embody the thoughts and feelings of the Creator as we know and feel them in ourselves intuitively.

Living Beyond Duality and Karma

When we desire to expand our awareness to our true potential, we become aware of our deep limitations that have kept us enclosed within the energetic spectrum of humanity. We have believed that we must depend upon others on many levels in order to survive. We have not known if our consciousness terminates at physical death. As a result, we live with some level of stress and fear. This is a state of negative polarity, and we occupy this space by our intention. By our intentional choice, we can decide to resolve all thoughts and emotions based in fear. We can realign ourselves with our natural state of positive polarity, with thoughts and feelings based in love, gratitude and joy.

This is a leap in consciousness. The realm of duality and karma has a negative polarity that is empowered by our own vibrations. It is based on fear and requires a positive polarity to dissolve from our experience into another energetic dimension. Without fear, we can know our true creative abilities. When we move energetically beyond the realm of karma, we are free to create anything we desire for ourselves and everyone around us. We are life-enhancers with our ability to be aware of all the energies around us and to choose to interface with positive, high-vibratory living.

By living in gratitude and being open to knowing the truth about ourselves, we become sensitive to our own inner knowing. Our intuition is always present as positive, high-frequency vibratory guidance that we can receive as knowing. This cannot be adequately described, but when we have it, we absolutely know it. Our intuition is our connection with the universal consciousness of the Creator. Coming to us in the conscious life force that we constantly receive, intuition unlimits our awareness and creative ability. We develop a perspective of compassionate wisdom.

The development of inner knowing depends primarily upon our desire and motivation. We all have it, but we have not known much of it. Once we are on the path to inner knowing, our life experiences become a kind of metaphor that guides our awareness to higher choices and states of being. All of our life becomes part of the process of life-enhancement through the energetic levels of love, compassion, joy and gratitude. We do not need anything from anyone outside of our own consciousness; although in our human form we enjoy providing for, and receiving from, one another. As we desire and feel ourselves living at the vibratory levels that we envision in all of our encounters, we radiate our energy into the quantum field for manifestation into our experiences. This is part of our creative ability.

As we become more and more positive in our awareness, we remove ourselves from the negative realm of karmic experiences. It's as if we're in parallel worlds. One is the realm of duality, of negative and positive energetics, and the other is beyond polarity and is pure life-enhancing creativity. While in the realm of duality, we arrive in the expanded realm by being positive. When we can align ourselves with positive polarity, we move into a higher dimension of energetics and living.

Activating Our Magnificence

As fractals of Creator consciousness, we have unlimited potential. In our human expression of being, we have limited ourselves within the empirical world of dense energetics and duality, and we've closed off our awareness from our true Being, but there is no requirement outside of ourselves and our limiting beliefs to remain in this condition. There are procedures that we can engage to expand ourselves greatly. Let's examine our own involvement in this realm.

Our primary tool in creation is our ability to focus our attention on any level of vibration that we can imagine and feel. We can choose our polarity—mentally and emotionally positive or negative. We were curious about fear, and so we experienced it. We've enslaved our consciousness to it, even believing in gradual life-diminishment as we age and die, with possible termination of our conscious being. This is the realm of negative polarity, which stimulates fear in us.

By intentionally being positive in every moment, we transform our lives and are able to do miraculous creations. We can talk with our Innate Being, who controls every cell in our body. It does not reason, it just follows directions and expresses the vitality of our body in alignment with our state of being. It listens when we acknowledge it and communicate lovingly with it. We can be grateful for the creation and vitality of our body, and we can forgive ourselves for messing it up with our ignorance and lack of loving attention. We can reach into the energy of our heart and be compassionate with ourselves. Our Innate can regenerate every cell instantly when we know this is true. In positive polarity, doubt does not exist.

Making the leap to positive in every moment requires strong intention and perseverance, along with developing sensitivity to intuition. When we realize positivity in our state of Being, we can communicate in confidence with the angels of the air, the Spirit of the Earth, and all conscious beings. We can ask the consciousness of the air to bring us the weather that we desire, and we can know the response intuitively. We can enter a higher energetic dimension by aligning with its vibrations. We can even disappear from the other dimension that we inhabit, and then reappear wherever we desire. It seems miraculous, but it's because we learn how to control our attention and awareness, directing our focus into positive, life-enhancing scenarios.

Our awareness expands into universal consciousness, which contains the empirical world, as long as human consciousness recognizes and feels its energetic patterns. We can include our body consciousness as we expand our awareness into a vibratory level of joy and gratitude. Our Innate loves to follow life-enhancing instructions of regeneration and increasing vitality. It is our natural state. We can communicate together for great benefit.

Intentionally Expanding Consciousness

We can ask our true, expanded Self to draw us into alignment with the consciousness of the Creator. We can seek the highest positive vibratory patterns that we can imagine. We can open our awareness beyond our imagination into universal consciousness. We ourselves have this ability, and we are the only force that can prevent us from realizing full consciousness. We must be willing and desirous of opening ourselves beyond the consciousness of humanity. For our ego-consciousness, opening to an unknown dimension of awareness stimulates fear.

It's comfortable to stay enclosed within our known human consciousness, but there is more, if we're interested. Within our common experience, we can choose what to think about, what to observe and how to feel about it all. If we are ready to expand, the way out of limitations occurs in our intention. Once we desire and pursue higher consciousness, our intuition guides and informs us. At first we stumble along, not sure of our inner knowing, but wanting to be confident in the truth in every aspect of life. With practice, intuitive knowing becomes our state of being.

We can become aware of greater depth and intensity in all aspects of life. This is the natural direction of our development.

We can choose to be positive and joyful. Nothing outside of ourselves can force us to feel anything. Every emotion is a choice, based on our perspective. Using our emotions creatively is part of our Being, and our perspective is an expression of our state of being. In every moment we control our perspective and state of being with the focus and feeling of our awareness.

We may encounter strong energetic patterns that stimulate emotions. Through our emotional intelligence, we can be aware of the quality of these energies, and we can decide if we want to engage them with a positive or negative perspective. If we choose to be in a life-enhancing state of being, we will engage our encounters with love, peace and compassion. When we remain in a highly positive state in the face of what could be a threatening situation, the situation resolves through transformation into alignment with us, or it dissolves into another dimension.

Consciously or unconsciously, we are the masters of our situation in every moment. We choose the quality of our experiences through our vibratory resonance. By intentionally being positive and joyful, we contribute to the enhancement of all conscious life. This is the path to greater awareness, far beyond compartmentalized human consciousness in the realm of duality.

Developing Our Multi-Dimensional Awareness

We have the ability to live in multiple dimensions simultaneously. We can choose which ones we want to pay attention to by aligning our vibrations with the ones we wish to engage with. Currently we are engaged in the empirical spectrum of dual polarities. Whenever we're motivated to move completely into positive polarity, we begin to align ourselves with the vibrations of love and compassion. As we go deeper into these emotions, we become aware of our true creative power residing in the essence of our consciousness. We have creative control of the level of vibrations of our state of being.

Chapter 6. Mastery of life

Whatever is happening outside of our own sense of being is unimportant in this quest. We can face our deepest fears and greatest doubt by recognizing their polarity, and instead being positive in a perspective of expansive awareness. Fear and doubt can exist for us only if we create them with our imagination and emotions. Our eternal presence of awareness does not need them, and we can let them go, reclaiming our life force from them and opening ourselves more completely to the unconditional love that is the essence of our Being.

Once we recognize the empirical spectrum as a compartment within our larger consciousness, we no longer are limited to the realm of duality in our awareness, but we can still focus in this realm as much as we want, while also being aware of our greater Being. We can shift our consciousness to awareness of a realm beyond polarity, the realm of infinite Being and unlimited creative ability and power in the consciousness of the Creator. This is our consciousness, once we resolve all of our limitations through our intuitive guidance.

Gaining confidence in our intuition can happen through being open and unencumbered mentally and emotionally. It may help to spend time in inspiring places in nature, being present in awareness and aligning with the energy of the Spirit of the Earth. We can realize that we have an inner knowing that guides us in every moment, and we can feel this inner level of vibration as joy. We can be in gratitude and feel enhancing energies.

When we're in a completely positive state of energetics, we're in a different dimension from that of duality, and the quality of our experiences becomes positive. Our expansiveness is beyond the imagination of our ego consciousness, requiring us to transcend the ego. This we can do by being sensitive to the positive, high-vibratory range of our intuition and willing to follow its guidance.

www.ingramcontent.com/pod-product-compliance
Lightning Source LLC
Chambersburg PA
CBHW020938180426
43194CB00038B/372